TEST PREP MATH BOOK
FOR CASAS MATH GOALS 2 LEVEL B

Helping Learners Approach Math with Confidence while Preparing them for CASAS Math GOALS 2 Level B—**Forms 923M and 924M**

By

TABLE OF CONTENT

PREFACE

Dear Instructors,

This Test Prep math book is specifically designed to prepare adult learners for the CASAS Math GOALS 2 Level B Forms 923M and 924M. It fully aligns with the CASAS Competencies and meets the requirements of the College and Career Reading Standards (CCRS), the National Reporting System (NRS), and the Workforce Innovation and Opportunity Act (WIOA).

Adhering to the CASAS test blueprint, this textbook covers mathematical areas through six detailed chapters: *Number Sense and Operations, Consumer Economics, Algebraic Thinking, Geometry, Data Analysis and Statistics, and Pure Mathematics.*

The book's content is structured to improve the mathematical thinking skills of adult students. It provides 14 lessons across various competencies, such as Consumer Economics, Community Resources, Employment, and Pure Mathematics. Each chapter is thoughtfully crafted with multiple lessons to foster deep understanding and practical application of mathematical concepts.

Additionally, the book includes two practice tests that simulate the actual CASAS assessments, incorporating real-world math problems to give students a genuine taste of what they can expect. Complete with answer keys for all exercises and practice tests, this textbook is a robust tool for effective learning and assessment.

Using this resource in your teaching will equip you to effectively develop and improve the math strategies, functions, and concepts necessary for your adult learners' success. Indeed, this book is an invaluable asset for programs that aim to empower their students with the mathematical skills required for success in everyday contexts such as community involvement, family management, and professional environments. To order class sets, go to cbledu.com.

INTRODUCTION

Dear Math Learners,

Welcome to your journey toward improving your math skills with this test-prep math textbook. It is designed specifically for adult learners like you. This book is structured to prepare you for the CASAS Math GOALS 2 Level B test. In the six chapters, you'll explore essential mathematical areas, including *Number Sense and Operations, Consumer Economics, Algebraic Thinking, Geometry, Data Analysis and Statistics, and Pure Mathematics*. With 14 practical lessons, this textbook offers a clear path to improving your mathematical knowledge.

Practicing the exercises in this book is crucial. Each chapter includes multiple lessons that build on each other to help you understand and apply mathematical concepts in real-world situations. By engaging with these exercises, you'll develop a stronger foundation in each topic, making sure that you're well-prepared not just for the tests but also for the practical application of these skills in daily life.

We've also included two practice tests that mimic the actual CASAS Level B test. These practice tests are designed to give you a realistic experience of what to expect on the actual test. By taking these practice tests, you can assess your progress, identify areas where you need further practice, and build your confidence.

This textbook is more than just a study guide—it's a tool that will equip you with the math strategies, functions, and concepts necessary to solve word problems confidently. Regular practice and study will transform your understanding of math, turning challenges into opportunities for growth and learning.

Remember, math skills are essential for success in various aspects of your life, including community involvement, managing family finances, and professional advancement. By committing to completing the exercises and fully engaging with the materials in this book, you'll be setting yourself up for success in your academic pursuits and beyond.

Let's get started on this path together!

HOW TO APPROACH MATH

Here are ten practical ways you can overcome math fear and anxiety and build confidence while using this math textbook:

1. **Start Small:** Begin with easier problems that you can solve to build your confidence before solving harder ones.

2. **Practice Regularly:** Consistent practice makes math feel more manageable. Try to work on math problems a few times a week.

3. **Use the Book's Resources:** Take advantage of the tools and explanations in your textbook. They are designed to help you understand and solve math problems.

4. **Take Breaks:** If you feel overwhelmed, take a short break. Come back to the problem with a clear mind.

5. **Ask for Help:** Don't hesitate to seek help when you need it. Ask a teacher, a classmate, or use online resources if you're stuck.

6. **Stay Positive:** Keep a positive attitude about math. Remind yourself that you can handle it and that it's okay to make mistakes as you learn.

7. **Set Small Goals:** Break your math studies into small, achievable goals. Celebrate when you reach these goals to motivate yourself.

8. **Understand, Don't Memorize:** Focus on understanding the math concepts rather than just memorizing formulas. This understanding will make you feel more confident in your ability to solve math problems.

9. **Visualize Success:** Picture yourself successfully solving problems and understanding concepts. This visualization can boost your confidence.

10. **Reflect on Progress:** Regularly look back at where you started and recognize the progress you've made. This can be a great confidence booster.

By following these strategies, you'll be better positioned to tackle math with less anxiety and more confidence.

STUDY STRATEGIES

Here are ten simple strategies to study and improve your math knowledge, skills, and understanding using this textbook. Each strategy is designed to be practical and straightforward.

Strategy	Description
1. **Set a study schedule.**	Allocate specific times each week for math study and practice to build a routine.
2. **Create a study space.**	Find a quiet, organized space dedicated to studying to stay focused.
3. **Use the textbook.**	Read explanations and solve problems in the textbook to understand concepts deeply.
4. **Practice with examples.**	Work through example problems to understand how to apply math rules before trying exercises on your own.
5. **Summarize each lesson.**	Write a brief summary of what you learned in each lesson to reinforce your understanding.
6. **Solve practice tests.**	Use practice tests in the textbook to prepare for the actual test and build confidence.
7. **Discuss with peers.**	Study in groups or discuss problems with classmates to get different perspectives and solutions.
8. **Teach someone else.**	Explain math concepts to someone else to improve your own understanding and retention.
9. **Use online resources.**	Supplement your textbook with online tutorials and exercises (e.g., YouTube videos) for additional practice.
10. **Review regularly.**	Regularly go back and review previous chapters to keep information fresh and build connections between topics and chapters.

Using these strategies can help you make the most of your math textbook and your study time to gradually build up your mathematical abilities and confidence.

CHAPTER 1:
NUMBER SENSE AND OPERATIONS

Lesson 1: Understand place value for whole numbers and decimals

Place Value System

In our decimal number system, the value of a digit depends on its place, or position, in the number. Each place has a value of 10 times the place to its right. A number in standard form is separated into groups of three digits using commas.

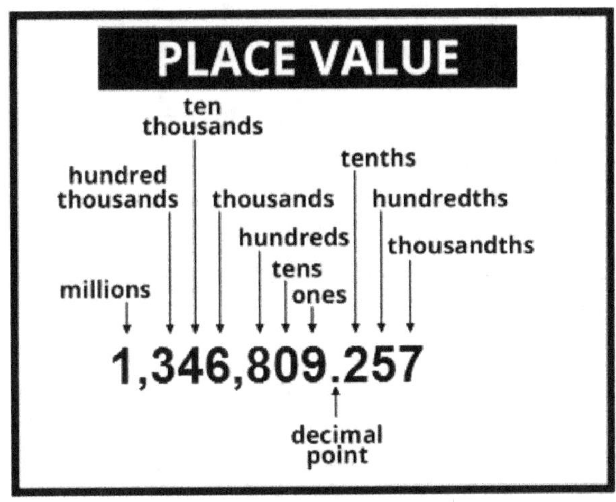

Numbers can be represented in several ways, but using the standard form is usually the easiest and shortest way.

Example 1:

What are some ways to show 3,146?

Solution:

Some ways to show 3,146 are:

In standard form: 3,146

As addition: $3,000 + 100 + 40 + 6$

In words: Three thousand one hundred forty-six

To read and write numbers, we use **base-ten numerals.** Base-ten numerals tell us that each place in the number is ten times the value of the place to its right. To write a number in words, first break down the number in expanded form, then combine the values into a number sentence.

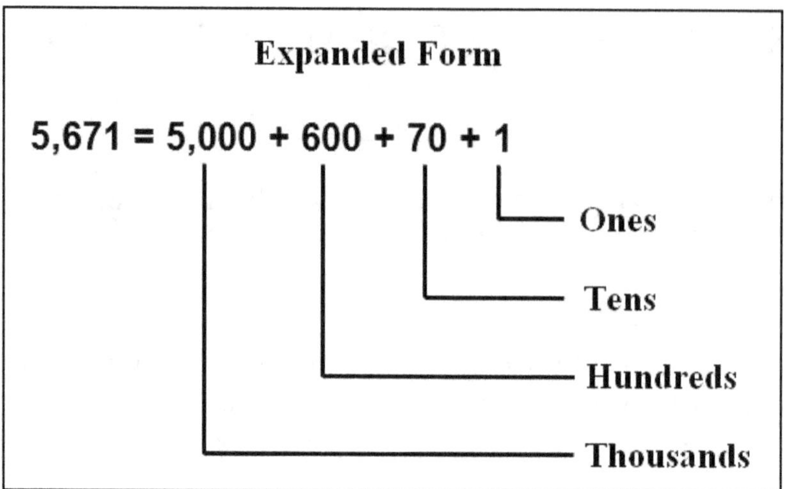

Example 2:

Write the number 7,539 in words.

Solution:

According to the place value, the first digit (**7**) is in the thousands place. This means there are **7 thousands** in the number 7,539.

The second digit (**5**) is in the hundreds place. This means there are **5 hundreds** in the number 7,539.

The third digit (**3**) is in the tens place. This means there are **3 tens** in the number 7,539.

The last digit (**9**) is in the ones place. This means there are **nine ones** in the number 7,539.

Thus, the number 7,539 in words is:

Seven thousand five hundred thirty-nine

Numbers expressed in the decimal form are called decimal numbers or **decimals.** A decimal has two parts: the whole number part and the decimal part. These parts are separated by a dot (**.**) called the **decimal point.**

To read or name decimal numbers:

1. Read the whole number part.

2. Say "and" to indicate the decimal point.

3. Read the decimal number.

4. Name **the decimal unit** where the last digit falls.

Example 2:

Write the number 3.184 in words.

Solution:

Read the whole part and the decimal part. Notice that the last digit, **4**, falls in the thousandth place.

Three and one hundred eighty-four thousandths

Practice Exercises

1. Write 9,581 in expanded form.

   A. 900 + 50 +81

   B. 9,000 + 50 + 8 + 1

   C. 9,000 + 500 + 80 + 1

   D. 9,000 + 500 + 8 + 10

2. What is the value of X?

$$11.589 = 10.000 + X + 500 + 80 + 9$$

   A. 1

   B. 1,100

   C. 11

   D. 1,000

3.

The estimated population of New York City is 8,335,897. What is the correct way to write this number?

   A. Eight hundred thirty-five thousand eight hundred ninety-seven

   B. Eight hundred thirty-three thousand eight hundred ninety-seven

   C. Eight million three hundred thirty-five thousand eight hundred ninety-seven

   D. Eight million three hundred thirty-three thousand eight hundred ninety-seven

4.

> A four-digit number has a 7 in the hundreds place, a 2 in the thousands place, a 5 in the tens place and an 8 in the ones place. What is the number?

A. 7,258

B. 2,785

C. 5,278

D. 2,758

> Chris wrote the following number in expanded form:
>
> $349,053 = 300,000 + A + 9,000 + B + 50 + 3$

5. What is the value of A?

A. 40,000

B. 4,000

C. 400

D. 4

6. What is the value of B?

A. 1

B.

C. 10

D. 100

E. 0

7. What is the correct way to write the number in words?

A. Three hundred forty-nine thousand fifty-three

B. Three hundred forty-nine thousand five hundred three

C. Three hundred forty-nine thousand thirty-five

D. Three hundred forty-nine thousand five

8. What is another way to show 101,563?

A. $100,000 + 10,000 + 500 + 63$

B. $110,000 + 1,000 + 500 + 60 + 3$

C. $100,000 + 1,000 + 500 + 60 + 3$

D. $100,000 + 1,000 + 50 + 60 + 3$

9. Which of the following numbers is five and forty-eight thousandths?

A. 5,048

B. 5.480

C. 5.840

D. 5.048

10. Which number shows a 7 in the hundredths?

    A. 706   C. 4.871

    B. 2.73   D. 70.75

Answer Key:

1) C	4) D	7) A	10) C
2) D	5) A	8) C	
3) C	6) D	9) D	

Lesson 2: Compute using the four operations

The four basic operations for all numbers in mathematics are:

$$\text{ADDITION } (+)$$
$$\text{SUBTRACTION } (-)$$
$$\text{MULTIPLICATION } (\times)$$
$$\text{DIVISION } (\div)$$

Addition is a process of adding things together. It is denoted by the plus sign (+). It involves combining two or more numbers into a single term. In addition, the order does not matter.

$$30 + 15 = 45$$

Subtraction shows the difference between two numbers. It is denoted by the minus sign (-). Subtraction is the inverse process of addition.

$$45 - 15 = 30$$

Multiplication is an operation that represents repeated addition of the same number. It is denoted by ($\times$). It combines two or more values to result in a single value.

$$6 \times 8 = 48$$

Division is an operation that involves sharing an amount into equal-sized groups. It is denoted by ($\div$). The division is the inverse process of multiplication.

$$48 \div 8 = 6$$

We can solve problems involving the four math operations.

Example 1:

There were 36 gallons of water in Jack's bathtub. Later, 15 gallons drained out. How much water is left in the bathtub?

Solution:

Notice that the problem requires the subtraction of whole numbers. Subtract the number of gallons.

$$36 - 15 = 21$$

Then, **21 gallons** of water are left.

Example 2:

Eddie has 6 gift cards to the same store. Each gift card is for $15. If he spends $30 of his gift card money today, how much will he have left?

Solution:

Step 1: First, find the total amount of gift card money by multiplying 6 by $15.

$$6 \times \$15 = \$90$$

Then, Eddie has $90 in gift card money.

Step 2: To find how much he will have left after spending $30, subtract $30 from $90.

$$\$90 - \$30 = \$60$$

Thus, Eddie will have **$60** left in gift card money.

Practice Exercises

1. Steve bought a smartphone for $245 and sold it for $358. How much profit did he make?

   A. $603   C. $113

   B. $124   D. $358

> Three books and an earphone cost $79. The earphone cost $28.

2. How much does each of the books cost?

   A. $51   C. $20

   B. $17   D. $15

3. What is the cost of 7 books?

   A. $119

   B. $357

   C. $196

   D. $553

Adam weighs 165 pounds. Mark is 8 pounds heavier than Adam. Kendrick is 12 pounds heavier than Mark.

4. How much does Mark weigh?

   A. 173 pounds

   B. 157 pounds

   C. 175 pounds

   D. 177 pounds

5. How much does Kendrick weigh?

   A. 169 pounds

   B. 177 pounds

   C. 180 pounds

   D. 185 pounds

6. How much heavier is Kendrick than Adam?

   A. 12 pounds

   B. 20 pounds

   C. 8 pounds

   D. 15 pounds

7. Ruth has 60 tickets for the fair, and each ride costs five tickets. How many rides can Ruth go on?

   A. 12

   B. 15

   C. 20

   D. 18

8. Ashley has 5 boxes with 9 cookies in each box. Joe has 4 boxes with 13 cookies in each box. How many more cookies does Joe have?

   A. 4

   B. 12

   C. 7

   D. 8

9. A movie theatre has 18 rows of seats, with 10 seats in each row. How many seats are there in total?

   A. 28

   B. 90

   C. 180

   D. 108

10. There are 14 doctors working in a clinic. Each doctor has 2 nurses assisting them. Four receptionists are working at the reception. How many people are working in the clinic?

A. 32

C. 46

B. 28

D. 50

Lesson 3: Perform operations with whole numbers, decimals, and fractions

To add and subtract whole numbers we must line up the numbers vertically by matching the **place values**, starting with the ones place.

Example 1:

The school library bought 495 new books last year. They bought 638 new books this year. How many new books did they buy in both years?

Solution:

Notice that we must add the numbers.

Step 1: First, line up the numbers vertically by matching the place values.

$$495 +$$
$$638$$

Step 2: Add the ones: $5 + 8 = 13$.

Step 3: When the sum of digits in a place value is greater than 9, you carry over the extra digit to the next higher place value. Since 13 is greater than 9, we write down the 3 in the ones place and carry over the 1 to the tens place.

$$1$$
$$495 +$$
$$638$$
$$3$$

Step 4: Add the tens, including the 1 that was carried over: $9 + 3 + 1 = 13$

$$11$$
$$495 +$$
$$638$$
$$33$$

Step 5: Add the hundreds, including the 1 hundred that was regrouped: $4 + 6 + 1 = 11$

$$\begin{array}{r} {}^{1\,1} \\ 495\;+ \\ 638 \\ \hline 1{,}133 \end{array}$$

Thus, they buy **1,133** new books.

Example 2:

Find the difference between 480 and 63.

Solution:

Step 1: To find the difference, we subtract 63 from 480. To do this, we line up the numbers vertically by matching the place values.

$$\begin{array}{r} 480\;- \\ 63 \\ \hline \end{array}$$

Step 2: Since we can't subtract 3 from 0 in the ones place, we need to borrow from the tens place. We must take 1 ten from the tens column in 480. We must give it to the zero in the ones column to make the number 10.

$$\begin{array}{r} {}^{10} \\ 480\;- \\ 63 \\ \hline \end{array}$$

Step 3: Calculate $10 - 3$ to get 7.

$$\begin{array}{r} {}^{10} \\ 480\;- \\ 63 \\ \hline 7 \end{array}$$

Step 4: Since we exchanged 1 ten from the tens column in 480, the number 8 becomes **7**.

$$\begin{array}{r} 7 \\ 4\!\!\not{8}0 \ - \\ \underline{63} \\ 7 \end{array}$$

Calculate 7− 6. We put the answer, 1, in the hundreds column.

$$\begin{array}{r} 7 \\ 4\!\!\not{8}0 \ - \\ \underline{63} \\ 417 \end{array}$$

Then, the result is **417**.

To multiply whole numbers, place the numbers so that one number is above the other and the numbers in the ones place are aligned.

Example 3:

Multiply 239 by 6.

Solution:

Step 1: Write the multiplier (6) under the multiplicand (239) and draw a line.

$$\begin{array}{r} 239 \ \times \\ \underline{6} \end{array}$$

Step 2: Multiply each digit of the multiplicand by the multiplier, starting from the right (ones place). Write the ones digit of each product below the line. If there is a tens digit, carry it and add it to the next product.

$$\begin{array}{r} 2\ 5 \\ 239 \ \times \\ \underline{6} \\ 1{,}434 \end{array}$$

6 times 9 is 54. Write 4 and carry 5.

6 times 3 is 18. Add 5 and it becomes 23. Write 3 and carry 2.

6 times 2 is 12. Add 2 and it becomes 14. Write 14.

Then, 239 x 6 = **1,434**

Example 4:

Calculate 932 x 47.

Solution:

Step 1: Align the multiplier (47) with the **ones** digit of the multiplicand (932), and draw a line.

$$
\begin{array}{r}
932 \ \times \\
\underline{47}
\end{array}
$$

Step 2: Multiply the multiplicand by each digit of the multiplier. Place the ones digit of each partial product in the same column as the multiplying digit.

$$
\begin{array}{r}
932 \ \times \\
\underline{47} \\
6524 \\
\underline{3728}
\end{array}
$$

932 x 7 = 6524

932 x 4 = 3728

Step 3: Add the partial products.

$$
\begin{array}{r}
932 \ \times \\
\underline{47} \\
6524 \ + \\
\underline{3728} \\
43,804
\end{array}
$$

Then, 932 x 47 = **43,804**

In division, the **dividend** is the number being divided while the **divisor** is the number we are dividing by.

To divide whole numbers, divide the first digit of the dividend by the divisor. If the divisor is larger than the first digit of the dividend, divide the first two digits of the dividend by the divisor, and so on. Write the quotient above the dividend.

Example 5:

Divide 3,174 by 6.

Solution:

Step 1: Rewrite the problem to set it up for long division.

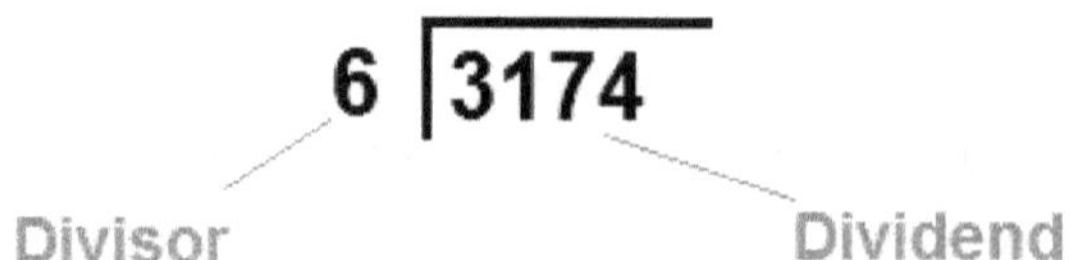

Step 2: Divide the first digit of the dividend, 3, by the divisor, 6.

$$6\,\overline{)3174}$$

Step 3: Since 6 does not go into 3, we use the first two digits of the dividend and divide 31 by 6. The divisor, 6, goes into 31 five times. We write the 5 in the quotient above the 1.

$$6\,\overline{)3\overset{5}{1}74}$$

Step 4: Multiply the 5 in the quotient by the divisor, 6, and write the product, 30, under the first two digits in the dividend.

$$\begin{array}{r} 5 \\ 6\,\overline{)3174} \\ 30 \end{array}$$

Step 5: Subtract that product from the first two digits in the dividend. Subtract $31 - 30$. Write the difference, 1, under the second digit in the dividend.

$$\begin{array}{r} 5 \\ 6\,\overline{)3174} \\ \underline{30} \\ 1 \end{array}$$

Step 6: Now bring down the 7 and repeat these steps. There are 2 sixes in 17. Write the 2 over the 7. Multiply the 6 by 2 and subtract the product (12) from 17.

$$
\begin{array}{r}
52 \\
6\,\overline{)3174} \\
30 \\
\overline{17} \\
12 \\
\overline{5}
\end{array}
$$

Step 7: Bring down the 4 and repeat these steps. There are 9 sixes in 54. Write the 9 over the 4. Multiply the 9 by 6 and subtract this product (54) from 54.

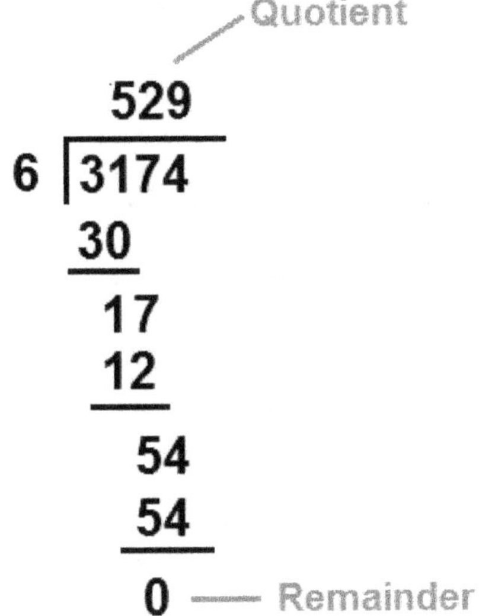

Thus, $3{,}174 \div 6 = \mathbf{529}$. We can check our answer by multiplying $529 \times 6 = 3{,}174$.

When adding and subtracting **decimals**, add or subtract as normal, but make sure you keep the decimal points aligned.

Example 6:

Calculate $18.37 + 2.09$

Solution:

Step 1: Keep the decimal points aligned and draw a line.

$$
\begin{array}{r}
18.37 \;\; + \\
2.09 \\
\hline
\end{array}
$$

Step 2: Add the numbers as whole numbers.

$$\begin{array}{r} {\scriptstyle 1 \quad 1} \\ \mathbf{18.37} \;\; \mathbf{+} \\ \underline{\mathbf{2.09}} \\ \mathbf{20.46} \end{array}$$

Then, 18.37 + 2.09 = **20.46**

To multiply **decimals**, ignore the decimal points (do not align them) and multiply the numbers as whole numbers. Then, starting from the right of the product, separate as many decimal digits as there are in the two numbers together.

Example 7:

Multiply 9.26 by 1.4

Solution:

Step 1: Ignore the decimal points and multiply the numbers as whole numbers.

$$926 \times 14 = 12964$$

Step 2: Now we must place the decimal point. Notice that 9.26 and 1.4 together have three decimal digits. Thus, the product will have three decimal digits.

Thus, 9.26 x 1.4 = **12.964**

To divide a decimal by another decimal number, we can multiply both the dividend and the divisor by 10 repeatedly until the divisor becomes a whole number.

Example 8:

Divide 3.6 by 0.08

Solution:

Multiply both numbers in the problem by 10 until the divisor is a whole number.

$$3.6 \div 0.08 \quad \text{(original problem)}$$

$$36 \div 0.8 \quad \text{(multiply by 10)}$$

$$360 \div 8 \quad \text{(multiply by 10; now the divisor is a whole number)}$$

The last problem, 360 ÷ 8, is equivalent to the original problem and easy to solve.

$$360 \div 8 = 45$$

Thus, 3.6 ÷ 0.08 = **45**

A **fraction** is a number that represents a part of a whole. For example, if we have a circle and we divide it into 4 equal slices, 1 of those slices is written as 1/4, as shown below:

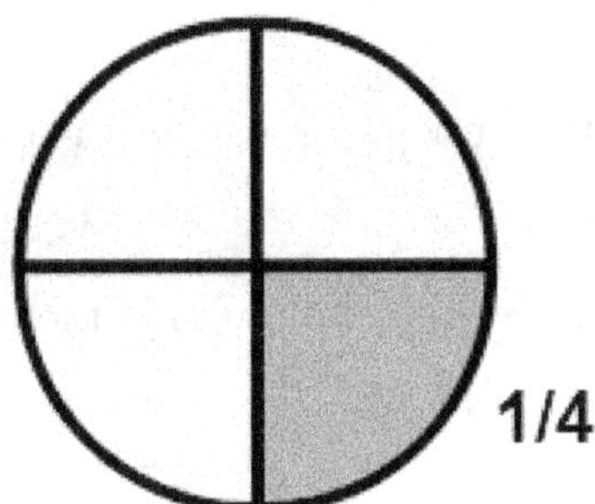

A fraction is made up of two parts: the denominator and the numerator.

The **denominator** is the bottom number of the fraction. It is the number of parts the whole is divided into.

The **numerator** is the top number of the fraction. It is the number of parts we have.

$$\frac{1}{4} \quad \Rightarrow \quad \textbf{Numerator}$$
$$\quad \Rightarrow \quad \textbf{Denominator}$$

To add and subtract **fractions**, we check to see if the fractions have the **same denominator**. If the fractions have the same denominator, add the top numbers (the numerators), and keep that same denominator.

Example 9:

Add $\frac{1}{5} + \frac{3}{5}$

Solution:

We check to see if the fractions have the same denominator. Since they do, we add the numerators and keep that same denominator.

$$\frac{1}{5} + \frac{3}{5} = \frac{4}{5}$$

If the fractions don't have the same denominator, then we must convert them to **equivalent fractions** with the same denominator. Once they have the same denominator, we can add or subtract the numbers in the numerator.

Example 10:

Calculate $\frac{1}{2} - \frac{1}{8}$

Solution:

Notice that the fractions don't have the same denominator, so convert them to equivalent fractions with the same denominator.

Step 1: To make the denominators the same, we must find a common denominator. The least common

denominator between 2 and 8 is 8. Therefore, we multiply the numerator and the denominator of the first fraction by 4 to get 8.

$$\frac{1 \times 4}{2 \times 4} = \frac{4}{8}$$

Step 2: Write the equivalent fraction. Since $\frac{1}{8}$ already has a denominator of 8, we can subtract the numerators and keep the same denominator.

$$\frac{4}{8} - \frac{1}{8} = \frac{3}{8}$$

So we get:

$$\frac{1}{2} - \frac{1}{8} = \mathbf{\frac{3}{8}}$$

To multiply fractions, we multiply the numerators of the fractions to get the new numerator, and multiply the denominators of the fractions to get the new denominator.

$$\frac{3}{5} \times \frac{4}{7} = \mathbf{\frac{12}{35}}$$

To divide fractions, we can follow these steps:

1. Turn the second fraction upside down. (This is now a **reciprocal.**)

2. Multiply the first fraction by that reciprocal.

3. Simplify the fraction (if needed).

Example 11:

Calculate $\frac{5}{6} \div \frac{10}{11}$

Solution:

Step 1: Turn the second fraction upside down. (It becomes a **reciprocal.**)

$$\frac{10}{11} \; becomes \; \frac{11}{10}$$

Step 2: Multiply the first fraction by that **reciprocal**:

$$\frac{5}{6} \times \frac{11}{10} = \frac{55}{60}$$

Step 3: Simplify the fraction. We do this by finding the **greatest common divisor** (GCD) of 55 and 60, which is 5. Then, divide the numerator and denominator by 5.

$$\frac{55 \div 5}{60 \div 5} = \frac{11}{12}$$

So we get:

$$\frac{5}{6} \div \frac{10}{11} = \mathbf{\frac{11}{12}}$$

Practice Exercises

1. Calculate $\frac{8}{11} - \frac{3}{11}$

   A. $\frac{11}{11}$

   B. $\frac{5}{11}$

   C. $\frac{1}{11}$

   D. $\frac{24}{11}$

2. Multiply 1.25 x 0.5

   A. 0.625

   B. 6.25

   C. 62.5

   D. 1.625

> On Friday, Macy's Supermarket sold 594 pounds of ground beef. On Sunday they sold three times that amount. On Monday they only sold 287 pounds.

3. How much meat did they sell on Sunday?

   A. 894 pounds

   B. 861 pounds

   C. 1,782 pounds

   D. 1,772 pounds

4. How much more meat did they sell on Sunday than Friday?

   A. 1,080 pounds

   B. 1,782 pounds

   C. 1,200 pounds

   D. 1,188 pounds

5. How much more meat did they sell on Sunday than Monday?

   A. 574 pounds

   B. 1,495 pounds

   C. 1,376 pounds

   D. 861 pounds

6. The value of Roger's college savings fund decreased by $1,875. If his fund was worth $5,000 before, how much is it worth now?

   A. $6,875

   B. $3,205

   C. $3,200

   D. $3,125

7. Which of the following is true?

   A. $30.50 + 7.50 = 38.00$

   B. $4.2 \times 4.2 = 1.764$

   C. $8.4 \div 0.3 = 2.8$

   D. $\frac{1}{5} + \frac{1}{3} = \frac{2}{8}$

8. There was $\frac{7}{8}$ of a pie left in the fridge. Nancy ate $\frac{3}{4}$ of the leftover pie. How much of the pie did she have?

   A. $\frac{1}{4}$

   B. $\frac{13}{8}$

   C. $\frac{1}{8}$

   D. $\frac{5}{6}$

9. A school is selling tickets at $8 each to attend the Summer Fair. In 12 weeks, it has earned $4,800. On average, how many tickets were sold each week?

   A. 40

   B. 50

   C. 65

   D. 96

10. It rained 3.43 inches on Saturday. On Monday, it rained 0.59 inches less than on Saturday. How much did it rain on Monday?

   A. 2.02 in.

   B. 4.02 in.

   C. 2.64 in.

   D. 2.84 in.

Answer Key:

1) B	4) D	7) A	10) D
2) A	5) B	8) C	
3) C	6) D	9) B	

Lesson 4: Understand ratio concepts and use ratios to solve problems.

A **ratio** is a way to show a relationship or compare two quantities of the same kind. There are different ways we write ratios, and they all mean the same thing. Here are some of the ways:

2 is to 3

2 : 3

$$\frac{2}{3}$$

A **rate** is a ratio of two quantities having different units, such as miles to hours. For example, if you buy 5 pounds of apples for $10, the rate is $10 for 5 pounds.

A **unit rate** is a rate where the second quantity is one unit. For example, if 5 pounds of apples cost $10, the unit rate would be $2 per pound. "Per pound" is the second quantity in the ratio.

Example 1:

Lucy is baking a cake using a recipe that calls for a butter-to-sugar ratio of 3:4. If she uses 16 cups of sugar, how many cups of butter should she use?

Solution:

Notice that Lucy uses four times as much sugar as the original recipe (4 x 4 = 16). Therefore, multiply the original butter amount by four as well.

$$3 \times 4 = 12$$

Notice that both ratios are equivalent:

$$\frac{3}{4} = \frac{12}{16}$$

Thus, Lucy needs **12 cups** of butter to maintain the 3:4 butter-to-sugar ratio.

Example 2:

A copy machine makes 150 copies in 50 seconds. Find the unit rate of copies per second.

Solution:

Divide 150 by 50 to find the unit rate.

$$\frac{150}{50} = 3$$

Then, the copy machine makes **3 copies per second.**

Practice Exercises

1. Morris bought 20 pounds of potatoes for $8. Find the unit rate.

   A. 2.5 dollars per pound

   B. 0.4 dollar

   C. 0.4 dollar per pound

   D. 4 pounds

2. A car can drive 450 miles on a tank of 20 gallons. How far can it drive on 60 gallons?

   A. 1,350 miles

   B. 900 miles

   C. 1,200 miles

   D. 2,700 miles

Look at the following table:

Item	Cost	Amount
A	$162	18 pounds
B	$144	12 pounds
C	$200	10 pounds

3. What is the unit rate of Item A in dollars per pound?

   A. $8 per pound

   B. $0.11 per pound

   C. $4.5 per pound

   D. $9 per pound

4. Which item has the smallest unit price?

   A. Item C

   B. Item A

   C. Item B

5. Which item has the greatest unit price?

   A. Item C

   B. Item B

   C. Item A

6. A wire costs $3 per foot. How much will 12 yards of wire cost? (Hint: 1 yard = 3 feet)

   A. $36

   B. $108

   C. $54

   D. $120

7. Which of the following is true?

   A. 118 cars / 3 weeks is a unit rate.

   B. 800 gallons / 50 minutes = 16 minutes per gallon

   C. 1,000 people per hour = 500 people per minute

   D. 360 bytes / 3 seconds = 120 bytes per second

8. Pete can run 5.5 miles in 0.5 hours. What is the unit rate in miles per hour?

   A. 10 miles per hour

   B. 11 miles per hour

   C. 9 miles per hour

   D. 10.5 miles per hour

The cost of 90 raffle tickets was $585.

9. What is the cost of one ticket?

   A. $9

   B. $8.50

   C. $10

   D. $6.50

10. What is the cost of 15 tickets?

   A. $97.50

   B. $39

   C. $6

   D. $80.50

Answer Key:

1) C	4) B	7) D	10) A
2) A	5) A	8) B	
3) D	6) B	9) D	

Answer the following reflection questions and feel free to discuss your responses with your teacher or a classmate.

1- What math ideas and principles did you learn in this chapter?

2- What new math concepts did you learn?

3- What procedures or methods did you practice in this chapter?

4- What aspect of this chapter is still not 100% clear for you?

5- What else do you want your teacher to know?

Lesson 1: Use measurement and money

We can measure how long or tall things are, or the distance between things, such as cities. These are all examples of length measurements. Length is typically measured in *metric units* or *customary units*. **Centimeters** and **meters** are examples of metric units.

Inches and **feet** are examples of customary units.

To measure lengths, we can use some tools (rulers and yardsticks).

A **foot** is longer than an **inch**. There are 12 inches in 1 foot.

1 foot = 12 inches

When 3 feet are together, this is called a yard.

1 yard = 3 feet

When we put together 1,760 yards, we have a mile.

1 mile = 1,760 yards

A **meter** is longer than a **centimeter.** There are 100 centimeters in 1 meter.

1 meter = 100 centimeters

Example 1:

Convert 20 yards to feet.

Solution:

We know that 1 yard = 3 feet. This means that the larger unit is a yard and the smaller unit is a foot.

To convert larger units (yards) to smaller units (feet), we **multiply** the number of larger units by the conversion factor, which is **3** in this case.

$$20 \text{ yards} = 20 \times 3 \text{ feet} = \textbf{60 feet}$$

Example 2:

Convert 9 centimeters to meters.

Solution:

Identify the larger unit and the smaller unit. The larger unit is the meter and the smaller unit is the centimeter. Since we are going from a larger unit to a smaller unit, we must **divide.**

Divide to find the number of meters in 9 centimeters:

$$9 \text{ cm} \div 100 = 0.09 \text{ m} \qquad (\text{Recall: } 1 \text{ m} = 100 \text{ cm})$$

Example 3:

Convert 3 kilometers to meters.

Solution:

Identify the larger unit and the smaller unit. The larger unit is the kilometer and the smaller unit is the meter. Since we are going from a larger unit to a smaller unit, we must **multiply.**

Multiply to find the number of meters in 5 kilometers:

$$3 \text{ km} \times 1000 = 3,000 \text{ m} \qquad (\text{Recall: } 1 \text{ km} = 1000 \text{ m})$$

Recall this:

Metric Units

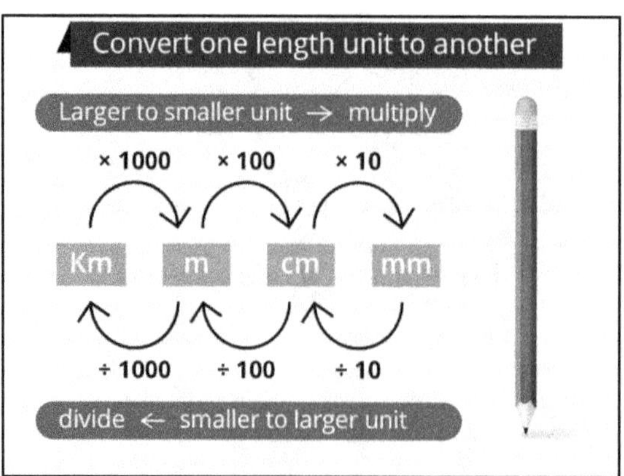

The smallest unit of mass is ounces (oz). A slice of pizza is about one ounce. If we have 16 ounces, it can also be called a pound (lb.). Typically, this is the unit that we use to measure people's weight.

1 pound = 16 ounces

When we put together 2,000 pounds, we have a ton. In other words, we have:

1 pound = 16 ounces

1 ton = 2,000 pounds = 32,000 ounces

Example 4:

Convert 92 pounds to ounces.

Solution:

Identify the larger unit and the smaller unit. The larger unit is a pound and the smaller unit is an ounce. Since we are going from a larger unit to a smaller unit, we must **multiply**.

Multiply to find the number of ounces in 92 pounds:

92 pounds x 16 ounces = 1,472 ounces   (Recall: 1 pound = 16 ounces)

People use money every day, and being able to count it and work out how much change we should be left with is an important life skill. The currency of the US is the United States Dollar (USD). Its symbol is $. American bills or paper currency comes in seven denominations: $1, $2, $5, $10, $20, $50, and $100.

The most commonly used coins in US money are quarters, dimes, nickels, and pennies.

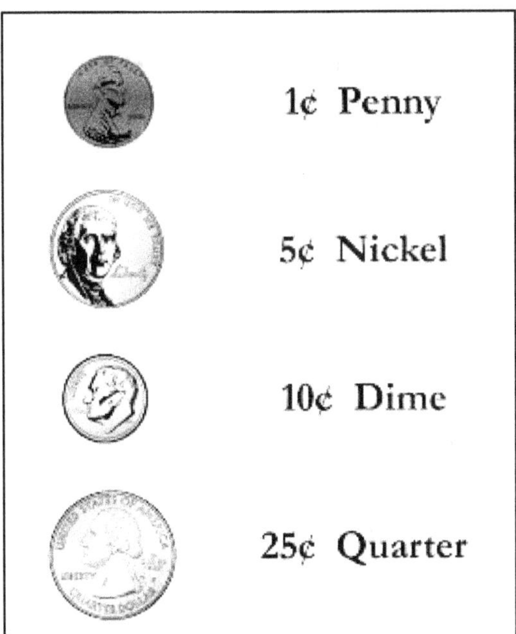

Some Dollar-Related Conversions:

1 dollar = 100 cents, so 1 cent is equal to 0.01 dollars.

1 nickel = 5 cents, so 1 nickel is equal to 0.05 dollars.

1 dime = 10 cents, so 1 dime is equal to 0.1 dollars.

1 quarter = 25 cents, so 1 quarter is equal to 0.25 dollars.

Example 5:

Arlene donated 750 cents and Darwin donated $12. In total, how many dollars did they donate?

Solution:

We know that 100 cents = 1 dollar. Then 750 cents = 750 ÷ 100 = 7.50 dollars. So Arlene donated $7.50. Thus, together they donated $7.50 + $12 = **$19.50**

Example 6:

What is the total value of 8 quarters, 2 $10 bills, 6 pennies, and 15 nickels?

Solution:

Add the coins and bills separately, and then add them together.

8 quarters = 8 x $0.25 = $2

15 nickels = 15 x $0.05 = $0.75

6 pennies = 6 x$0.01 = $0.06

2 $10 bill = 2 x $10 = $20

The coins are worth $2.81. With the 2 $5 bill, that's $2.81 + $20 = **$22.81**

Practice Exercises

Michael rode 5 miles on his bike. Karen rode 7,040 yards on her bike, and Dave rode 10,560 feet on his bike.

1. How many miles did Karen ride?

A. 4 miles

B. 5 miles

C. 7 miles

D. 4.5 miles

2. How many feet did Michael ride?

A. 26,400 ft.

C. 13,200 ft.

B. 8,800 ft.

D. 15,000 ft.

3. How many yards did Michael ride?

A. 26,400 yd.

C. 7,500 yd.

B. 13,200 yd.

D. 8,800 yd.

4. Who rode the farthest?

A. Dave

C. Karen

B. Michael

5. Bruce weighs 110 pounds. What is Bruce's weight in ounces?

A. 1,110 ounces

C. 1,760 ounces

B. 2,000 ounces

D. 880 ounces

6. Which of the following is the smallest?

A. 3 feet

C. 30 inches

B. 1 yard

D. 0.5 miles

7. Judith has 10 dollar bills, 8 quarters, and 12 dimes. How much money does she have?

A. $13.20

C. $30.00

B. $13.00

D. $13.50

8. Which of the following is true?

A. $0.01 = 1 penny

C. 10 dimes = 5 quarters

B. 10 quarters = $1

D. 100 pennies = $10

9. Terry claims there are 4,000 meters in X kilometers. What is X?

A. 40

C. 400

B. 4

D. 44

10. Caitlin claims there are Y quarters in $4.50. What is Y?

    A. 16

    B. 18

                                               C. 20

                                             D. 15

Answer Key:

1) A	4) B	7) A	10) B
2) A	5) C	8) A	
3) D	6) C	9) B	

Lesson 2: Use information to identify and purchase goods and services.

A good is a physical item that can be bought, touched, and used. A service is the action done for people who pay for the service.

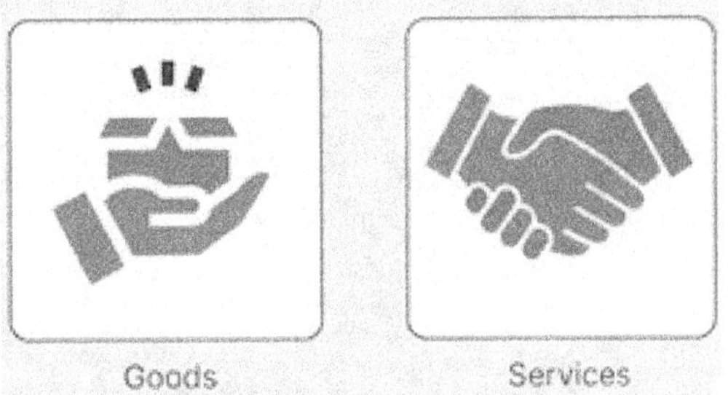

To obtain the best deal, we have to compare the cost of two or more items and then decide which is the best value.

To compare prices, divide the cost by the weight or quantity of the item. Then, we can compare the two items. In other words, to compare prices, we compare the **unit rates** of the items.

Example 1:

At Mega Store, Brenda could get 3 T-shirts for $27. At an online shop, the price for 6 T-shirts is $51. Which is the best deal?

Solution:

Find the unit price for each place:

Mega Store:

$$unit\ rate = \frac{\$27}{3} = \$9\ per\ T-shirt$$

Online Store:

$$unit\ rate = \frac{\$51}{6} = \$8.50\ per\ T-shirt$$

Then, the best deal is 6 T-shirts for $51 (lower unit rate).

Reading bills and receipts is a life skill because almost every household receives at least one type of bill or receipt each month. Understanding how to read bills and receipts can help us budget our money.

Example 2:

According to the following gas station receipt, what is the cost of a gallon of fuel?

```
          FUEL DEPOT
       1 GOODSPRINGS RD.
         JEAN, NV 89019
          702-761-7000
    ------------------------------
    DATE            07/10/2020
    Time            10:40 AM
    PUMP            8
    TRAN#           171
    ------------------------------
                 DETAILS
    BASE PRICE   $ 2.97 / GAL
    GALLONS        33.1820
    TOTAL        $ 98.55
    ------------------------------
    $ 98.55 REG FUEL
    $ 4.43 TAX
    $-102.98 VISA DEBIT PAID

    $0.00 BALANCE


       THANK YOU FOR VISITING
            FUEL DEPOT
```

Solution:

Notice that the base price is $2.97/gal. This means the unit price of fuel is $2.97 per gallon. Then, the cost of a gallon of fuel is **$2.97**

Practice Exercises

Look at the following receipt:

```
       Seaside Sushi House
          1500 Main Ave
       Long Beach, CA 90712
          505-303-2993

    ****************************
    09/09/2020         06:45 AM
             TERMINAL 1
    ****************************
    1    Rainbow Roll       $15.95
    1    Spider Roll        $14.95
    1    750ml Hakutsuru    $39.95

         SUB-TOTAL        $70.85
         TAX               $5.31

    PAYMENT TYPE        VISA Card
    APP# : 11278860
    REF# : 18623058
    REC# : 0018

       TOTAL DUE          ?
```

1. Which item is the most expensive?

A. Spider Roll

C. Rainbow Roll

B. Hakutsuru

2. What is the sub-total before tax?

A. $5.31

C. $39.95

B. $70.85

D. $75.00

3. What is the total amount?

A. $70.85

C. $71.00

B. $65.64

D. $76.16

4. What is the cost of three Spider Rolls?

A. $47.85

C. $44.85

B. $45.95

D. $119.85

A pizza shop has three different offers.

Offer 1	Offer 2	Offer 3
3 pizzas	6 pizzas	12 pizzas
$27	$48	$120

5. What is the unit price of Offer 1?

A. $9 per pizza

C. $5 per pizza

B. 9 pizzas per dollar

D. 0.9 pizzas per dollar

6. What is the best deal?

A. Offer 3

C. Offer 2

B. Offer 1

7. If Rose chooses Offer 2, what is the cost of 12 pizzas?

A. $102

C. $120

B. $108

D. $96

Look at the following receipt:

8. How many items were purchased?

    A. 3   C. 1

    B. 4   D. 5

9. Which item is the cheapest?

    A. Pain Killer   C. Fever tabs

    B. Flu medicine

10. What is the sales tax?

    A. $1.74   C. $2.50

    B. $23.50   D. $1.50

Answer Key:

1) B	4) C	7) D	10) A
2) B	5) A	8) B	
3) D	6) C	9) C	

Answer the following reflection questions and feel free to discuss your responses with your teacher or a classmate.

1- What math ideas and principles did you learn in this chapter?

2- What new math concepts did you learn?

3- What procedures or methods did you practice in this chapter?

4- What aspect of this chapter is still not 100% clear for you?

5- What else do you want your teacher to know?

CHAPTER 3: ALGEBRAIC THINKING

Lesson 1: Apply properties of the four operations.

Commutative Property

Commutative property tells us that in some mathematical operations, it does not matter if the terms to operate are placed in one order or another.

> ### *Commutative Property of Addition*
>
> ### A + B = B + A
>
> "Changing the order of addends does not change the sum."

> ### *Commutative Property of Multiplication*
>
> ### A x B = B x A
>
> "The order of the factors does not alter the product."

Notice that this property does not apply **to subtraction and division.**

$$15 - 5 = 5 - 15$$
$$8 \div 4 = 4 \div 8$$

Associative Property

When a math operation is very long and has many terms, we can solve it by grouping the terms.

> ### *Associative Property of Addition*
>
> ### A + B + C = (A + B) + C = A + (B + C)
>
> "We can group addends in any order, and the sum will be the same."

> ### *Associative Property of Multiplication*
>
> **A x B x C = (A x B) x C = A x (B x C)**
>
> "We can group the factors in different ways, and the product will be the same."

Notice that this property does not apply **to subtraction and division.**

Distributive Property

According to the distributive property, multiplying the sum or difference of two or more terms by a number will give the same result as multiplying each term individually by the number and then adding or subtracting the products together.

In other words, the distributive property describes how we can distribute multiplication over addition and subtraction.

> ### *Distributive Property of Multiplication*
>
> **A (B + C) = AB + AC**
>
> **A (B − C) = AB − AC**

We can use this property for division.

> ### *Distributive Property of Division*
>
> **(B + C) ÷ A = (B ÷ A) + (C ÷ A)**
>
> **(B − C) ÷ A = (B ÷ A) − (C ÷ A)**

Example 1:

Find the sum of 39 + 43 + 51 + 27

Solution:

We can use the **associative property of addition** to group the addends in a different order.

$$(39 + 51) + (43 + 27)$$

$$90 + 70 = \mathbf{160}$$

Notice that we obtain easy sums.

Example 2:

Find the product of 8 x 4 x 35 x 25

Solution:

We can use the **commutative property of multiplication** to make the operations easier to carry out.

$$(8 \text{ x } 35) \text{ x } (4 \text{ x } 25)$$

$$280 \text{ x } 100 = \mathbf{28{,}000}$$

Example 3:

Find the value of 6 (23 – 15)

Solution:

We can use the distributive property of multiplication:

$$6 \ (23 - 15) = 6 \text{ x } 23 - 6 \text{ x } 15$$

$$6 \ (23 - 15) = 138 - 90 = 48$$

Practice Exercises

1. Which of the following is equivalent to 37 + 18 + 23?

    A. 18 + 37 + 23

    B. 18 x 23 + 37

    C. 18 (37 + 23)

    D. 37 (23 – 18)

2. What is the value of M?

$$14 \text{ x } 59 = 59 \text{ x } M$$

    A. 59

    B. 41

    C. 28

    D. 14

The yield of an apple farm was 345 dozen apples.

3. Which expression represents the total number of apples harvested?

   A. 345 + 12

   B. 300 + 45 + 12

   C. 345 (10 + 2)

   D. 345 (10 – 2)

4. How many apples were harvested?

   A. 2,760

   B. 4,140

   C. 4,200

   D. 3,500

5. What is the value of Z?

$$32 + 5 + 19 + 112 = 5 + 32 + Z + 19$$

   A. 112

   B. 100

   C. 110

   D. 89

6. The expression 60 (15 – 6) equals:

   A. 360 – 900

   B. 615 – 360

   C. 900 + 360

   D. 900 – 360

7. Which of the following is true?

   A. $45 - 19 = 19 - 45$

   B. $3 + 6 + 9 = 3 - 6 - 9$

   C. $(45 - 9) \div 3 = 15 - 3$

   D. $10 \times 11 \times 12 = 12 (11 + 10)$

8. Which expression is equivalent to 79 x 104?

   A. 104 + 70 + 9

   B. 79 (100 + 4)

   C. 79 (100 – 4)

   D. 104 ÷ (70 + 9)

9. What is the value of N?

$$5 (6 + 30) = 30 + N$$

A. 15

B. 35

C. 150

D. 180

10. What is the missing value?

$$M + N + P + Q = (N + ?) + (Q + M)$$

A. P

B. Q

C. 2P

D. 3N

Answer Key:

1) A

2) D

3) C

4) B

5) A

6) D

7) C

8) B

9) C

10) A

Lesson 2: Use a symbol to represent variables and solve simple one-variable equations.

A **variable** is a symbol for a number we don't know yet. It is usually a letter like x or y, but we can use any letter. An **equation** is a mathematical statement that means two things are equal.

In an equation, the left side is always equal to the right. The most common equations contain one or more **variables**.

$$2x - 3 = 18$$

To solve a simple one-variable equation, follow these steps:

1) Figure out what to remove to get the value of the variable.

2) To remove a number, add its opposite to both sides.

We can solve word problems using equations with variables for unknown numbers to represent the problem or simple contextual math situations.

Example 1:

The cost of a table and a chair is \$101. If the table costs \$55 more than the chair, find the cost of the table and the chair.

Solution:

Step 1: Let x be the cost of the chair. Then, the cost of the table is x + 55.

Step 2: Set up the equation that represents the problem:

$$x + x + 55 = 101$$

Step 3: Solve the equation. Combine like terms:

$$\mathbf{x + x} + 55 = 101$$

$$\mathbf{2x} + 55 = 101$$

Step 4: We want to remove 55 in the equation. To remove 55, do the opposite. In this case, subtract 55 from both sides of the equation.

$$2x + 55 - 55 = 101 - 55$$

$$2x = 46$$

Step 5: We want to remove 2 in the equation. To remove 2, do the opposite. In this case, divide both sides of the equation by 2

$$\frac{2x}{2} = \frac{46}{2}$$

$$x = 23$$

Thus, the cost of the chair is **$23** and the cost of the table is **$78**.

Example 2:

Diane and Lucy submitted a math test. They scored 128 points. Lucy scored three times more points than Diane. How many points did Diane score?

Solution:

Step 1: Let x be Diane's score. Then, Lucy's score is 3x. Set up the equation that represents the problem.

$$3x + x = 128$$

Step 2: Solve the equation. Combine like terms:

$$\mathbf{3x + x = 128}$$

$$\mathbf{4x = 128}$$

Step 3: We want to isolate x so we must remove 4 from the equation. To remove 4, do the opposite. In this case, divide both sides of the equation by 4.

$$\frac{4x}{4} = \frac{128}{4}$$

$$x = \mathbf{32}$$

Thus, Diane scored **32 points.**

Practice Exercises

1. Solve the following equation:

$$\mathbf{2x - 12 = 20}$$

A. x = 4

B. x = 32

C. x = 24

D. x = 16

> Joe has several books in his house. He gives 38 books to his friends, and he keeps the remaining 19 books.

2. Which equation can we use to find the total number of books (x) in Joe's house?

   A. $x - 38 = 19$

   B. $x + 38 = 19$

   C. $x - 19 = 38$

   D. $38x = 19$

3. What is the total number of books in Joe´s house?

   A. 19

   B. 57

   C. 46

   D. 55

> Laura went to the zoo with her family. The cost was $12 for parking plus $15 per person to enter the zoo. Laura and her family spent $102. Let n be the number of people who entered the zoo.

4. Which equation represents the problem?

   A. $n + 12 + 15 = 102$

   B. $15n + 12 = 102$

   C. $12n + 15 = 102$

   D. $12n + 15n = 102$

5. How many people entered the zoo?

   A. 6

   B. 5

   C. 8

   D. 9

6. Four times a number is equal to 148. What is the number?

   A. 35

   B. 144

   C. 37

   D. 43

7. If $0.8y - 2.1 = 1.9$, what is y?

   A. 3.2

   B. 6.4

   C. 4

   D. 5

8. Which of the following is the solution of the equation $10x - 1 = 0$?

   A. $x = 1$

   B. $x = 10$

   C. $x = 0.1$

   D. $x = 0.5$

9. Twice a number decreased by 52 is equal to 50. What is the number?

    A. 102   C. 52

    B. 51   D. 45

10. $1,835 is shared between Byron and Darryl. Byron's share is $565 less than Darryl's share. What is Byron's share?

    A. $1,000   C. $600

    B. $635   D. $1,200

Answer Key:

1) D	4) B	7) D	10) B
2) A	5) A	8) C	
3) B	6) C	9) B	

REFLECTION ON LEARNING

Answer the following reflection questions and feel free to discuss your responses with your teacher or a classmate.

1- What math ideas and principles did you learn in this chapter?

2- What new math concepts did you learn?

3- What procedures or methods did you practice in this chapter?

4- What aspect of this chapter is still not 100% clear for you?

5- What else do you want your teacher to know?

CHAPTER 4:
GEOMETRY

Lesson 1: Solve perimeter and area problems.

Perimeter is the distance around a two-dimensional shape.

Area is the amount of two-dimensional space inside a closed two-dimensional figure. A square with side length 1 unit, called a **unit square**, is said to have "one square unit" of area, and can be used to measure area. A square inch is a unit of area equal to the area of a square with sides of one inch.

The Perimeter and Area of Common Two-Dimensional Shapes

		Perimeter	Area
Rectangle	Length (l), Width (w)	$P = 2(l + w)$	$A = l \times w$
Square	Length (x), Width (x)	$P = 4x$	$A = x^2$
Parallelogram	a, h, b	$P = 2(a + b)$	$A = a \times h$
Trapezoid	a, c, h, d, b	$P = a + b + c + d$	$A = \dfrac{(a + b) \times h}{2}$
Triangle	c, h, a, b	$P = a + b + c$	$A = \dfrac{b \times h}{2}$
Circle	r	$P = Circumference = 2\pi r \text{ or } \pi D$	$A = \pi \times r^2$

Example 1:

Find the area of a triangle with a base of 12 inches and a height of 8 inches.

Solution:

Apply the formula of the area of a triangle.

$$Area = \frac{b \times h}{2}$$

Replace the values in the formula:

$$Area = \frac{12\ in \times 8\ in}{2} = \frac{96\ in^2}{2} = \mathbf{48\ in^2}$$

Example 2:

Find the perimeter of the following rectangle:

Solution:

Notice the dimensions of the rectangle: width = 8 ft. and length = 20 ft.

Apply the formula of the perimeter of a rectangle:

$$P = 2 \cdot (l + w)$$

$$P = 2 \cdot (20\ ft. + 8\ ft.) = 2 \cdot (28\ ft.) = \mathbf{56\ ft.}$$

Example 3:

The area of a square is 81 square feet. What is the length of the side of the square?

Step 1: Apply the formula of the area of a square.

$$A = l^2$$

Step 2: Where A = 81 ft², apply the value in the formula and we get an equation:

$$l^2 = 81\ ft^2$$

Step 3: Find the square root of both sides:

$$\sqrt{l^2} = \sqrt{81\ ft^2}$$

$$l = 9\ ft$$

Thus, the length of the side of the square is **9 feet**.

Practice Exercises

The width of a photo frame is 8 inches. The height is three times the width.

1. What is the height of the photo frame?

   A. 11 in.   C. 24 in.

   B. 18 in.   D. 24 in^2

2. What is the perimeter of the photo frame?

   A. 32 in.   C. 22 in.

   B. 64 in.   D. 48 in.

3. What is the area of the photo frame?

   A. 192 in^2 C. 200 in^2

   B. 128 in^2 D. 240 in^2

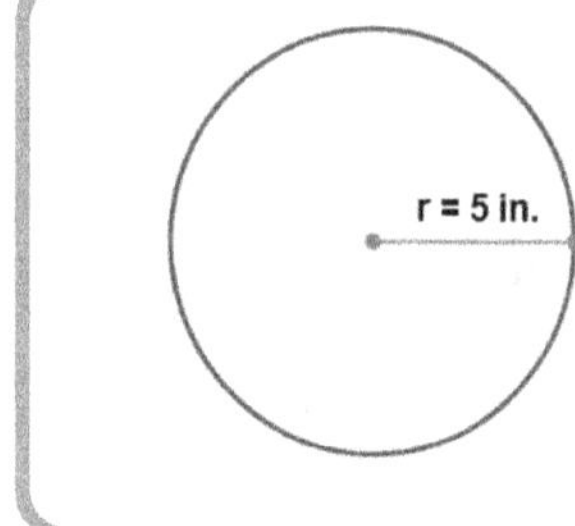

The radius of a circle is 5 inches. The diameter of a circle is twice the radius.

4. What is the diameter of the circle?

   A. 7 in.   C. 8 in.

   B. 25 in.   D. 10 in.

5. What is the circumference of the circle? (Use $\pi = 3.14$)

   A. 314 in.

   B. 31.4 in.

   C. 15.7 in.

   D. 35 in.

6. What is the area of the circle? (Use $\pi = 3.14$)

   A. 314 in^2

   B. 157 in^2

   C. 31.4 in^2

   D. 78.5 in^2

7. What is the area of the following shape?

15 ft

13 ft.

21 ft

   A. 468 ft^2

   B. 273 ft^2

   C. 234 ft^2

   D. 315 ft^2

8. What is the base of a triangle whose height is 20 inches and area is 150 square inches?

   A. 15 inches

   B. 30 inches

   C. 7.5 inches

   D. 25 inches

9. If the perimeter of a square is 20 feet, what is the area of the square?

   A. 10 ft^2

   B. 20 ft^2

   C. 40 ft^2

   D. 25 ft^2

10. The lengths of the sides of a triangle are 9 inches, 12 inches and x inches. The perimeter of the triangle is 36 inches. What is x?

   A. 21

   B. 16

   C. 15

   D. 18

Answer Key:

1) C	4) D	7) C	10) C
2) B	5) B	8) A	
3) A	6) D	9) D	

Lesson 2: Measure with non-standard and metric units and convert within a given measurement system.

Metric Units

Small metric units of length are called **millimeters.** The symbol for millimeters is **mm.** When we have something that is 10 millimeters long, it is said to be 1 **centimeter**. The symbol for centimeters is **cm.** In other words,

1 centimeter = 10 millimeters

1 cm = 10 mm

A **meter** is longer than a **centimeter.** There are 100 centimeters in 1 meter.

1 meter = 100 centimeters

In other words,

1 meter = 100 centimeters = 1,000 millimeters

1 m = 100 cm = 1,000 mm

When we need to measure greater distances, such as how far we are going to drive or fly between cities, we measure that distance using miles or **kilometers.**

1 km = 1,000 m

The two most common measurements of volume are milliliters (ml) and liters (l). A **milliliter** is a very small amount of liquid. For example, a teaspoon can hold about five milliliters.

A **liter** is made of many milliliters. Specifically, 1,000 milliliters make up 1 liter:

1 liter = 1,000 milliliters

When we buy groceries, they are measured in weight units such as grams and kilograms. A **gram** is a metric unit of mass (or weight). It is abbreviated as **g**. A **kilogram** is another measure of mass (or weight). It is abbreviated as **kg**.

$$1 \text{ kilogram} = 1,000 \text{ grams}$$

$$1 \text{ kg} = 1,000 \text{ g}$$

Non-Standard Units

A non-standard unit of measure is a specified amount that is used repeatedly to indicate the length, volume or mass of an object. Non-standard units are everyday items or objects that can be used to estimate the length, weight, or volume of an object.

Examples of non-standard units of measure:

Length– paper clips, pencils, hand span, cubit or foot span

Weight– apples, books, boxes or toy blocks

Volume– cups, spoons, handfuls or barrel

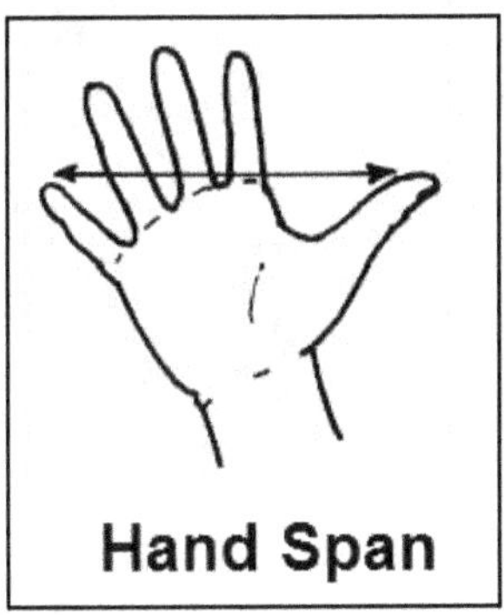

We can convert non-standard units to metric units using the following relationships:

$$1 \text{ hand span} = 22.86 \text{ centimeters}$$

$$1 \text{ cup} = 237 \text{ milliliters}$$

$$1 \text{ spoon} = 14.79 \text{ milliliters}$$

$$1 \text{ barrel} = 158 \text{ liters}$$

Example 1:

Convert 5 hand spans to centimeters.

Solution:

Identify the larger unit and the smaller unit. The larger unit is the hand span and the smaller unit is the centimeter. Since we are going from a larger unit to a smaller unit, we must **multiply.**

Multiply to find the number of centimeters in 5 hand spans:

$$5 \times 22.86 \text{ cm} = \textbf{114.3 cm}$$

Practice Exercises

Look at the following weight scale:

1. Which of the following is true?

    A. Ten cans are heavier than the box.

    B. Ten cans weigh the same as the box.

    C. The height of the box is equal to the height of the ten cans.

    D. Ten cans is a metric unit of length.

2. Suppose that the weight of one can is 250 grams. What is the weight of the box in kilograms?

    A. 2,500kg

    B. 2.5 kg

    C. 25 kg

    D. 250 kg

3. Suppose that the weight of the box is 3 kilograms. What is the weight of a can in grams?

    A. 300 g

    B. 30 g

    C. 3 g

    D. 3,000 g

4. Suppose that the weight of the box is 4,500 grams. What is the weight of the ten cans?

   A. 450 g

   B. 450 kg

   C. 45 g

   D. 4,500 g

Luke wants to measure the height of a building using a light pole as shown below:

5. Which of the following is true?

   A. The height of the building is about five times the height of the light pole.

   B. The height of the building is about two times the height of the light pole.

   C. The height of the building is about three times the height of the light pole.

   D. The height of the building is a metric unit of length.

6. If the height of the light pole is 7.50 meters, approximately how tall is the building?

   A. 10.50 m

   B. 15 m

   C. 30 m

   D. 22.50 m

7. If the height of the building is 33 meters, approximately how tall is the light pole?

   A. 11 m

   B. 3 m

   C. 8.50 m

   D. 12 m

8.

 What unit would we use to find the weight of an iPhone?

A. Kilograms

C. Grams

B. Cups

D. Hand span

9. What unit would we use to measure the capacity of a bathtub?

A. Spoons

C. Handfuls

B. Liters

D. Milliliters

10. The height of a rack is 13 hand spans. What is the height of the rack in meters?

A. 2.97 m

C. 29.70 m

B. 297 m

D. 3.50 m

Answer Key:

1) B	4) D	7) A	10) A
2) B	5) C	8) C	
3) A	6) D	9) B	

Lesson 3: Solve measurement word problems, including those with time and volume.

To measure the time, we use **hours, minutes, and seconds**.

For example, if the time is **9:25,** the hour is **9** and the minutes are **25**. Another way to say this is that the time is 25 minutes after 9 o'clock.

The basic units of time that we are familiar with from the smallest unit to the greatest unit are **second, minute, hour, day, week, month, and year**. The second is the smallest unit of time. These units have the following relations between each other:

1 minute = 60 seconds

1 hour = 60 minutes

1 day = 24 hours

1 week = 7 days

1 year = 12 months = 365 days

(Note: In a leap year, the number of days is 366.)

The U.S. customary capacity or volume measurement units are ounces, cups, pints, quarts, and gallons. The smallest unit of capacity is ounces (or fluid ounces). A cup is equal to 8 ounces.

1 cup = 8 fluid ounces

When we put together 2 cups, we have a pint.

1 pint = 2 cups

A quart (qt) is the same thing as 4 cups or 2 pints.

1 quart = 2 pints = 4 cups

The largest unit of capacity is gallon (gal). A gallon is the same as 16 cups or 8 pints or 4 quarts.

1 gallon = 4 quarts = 8 pints = 16 cups

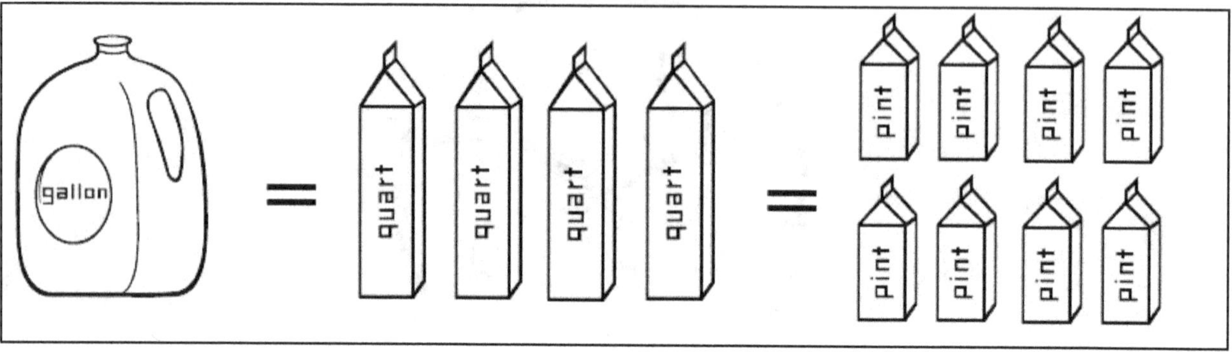

Example 1:

How many days are 264 hours?

Solution:

To convert smaller units (hours) to larger units (days), we **divide** the number of smaller units by **24.** Recall: 1 day = 24 hours.

$$264 \text{ hours} = \frac{264}{24} \text{ days} = \textbf{11 days}$$

Example 2:

How many quarts are there in 23 gallons?

Solution:

Identify the larger unit and the smaller unit. The larger unit is a gallon and the smaller unit is a quart.

Since we are going from a larger unit to a smaller unit, we must **multiply.** Multiply to find the number of quarts in 23 gallons:

23 x 4 = **92 quarts**

(Recall: 1 gallon = 4 quarts)

Practice Exercises

> A race took Pete 5 minutes and 40 seconds to complete. The race took Janice a quarter of an hour. The same race took Luke 960 seconds.

1. What was Janice's time in seconds?

   A. 3,600 seconds

   B. 1,000 seconds

   C. 950 seconds

   D. 900 seconds

2. What was Luke's time in minutes?

   A. 15 minutes

   B. 30 minutes

   C. 16 minutes

   D. 18 minutes

3. Who was the fastest?

   A. Janice

   B. Pete

   C. Luke

   D. Cannot be determined

4. Which of the following is true?

   A. 5 years = 60 months

   B. 1 gallon = 6 quarts

   C. 1.5 hours = 100 minutes

   D. 300 seconds = 3 minutes

5. It takes Charlie 17 minutes to bike to school. If he walks, it takes him twice as long. Charlie leaves for school at 8:30 a.m. What time will he get to school if he walks?

   A. 8:54 a.m.

   B. 9:04 a.m.

   C. 8:47 a.m.

   D. 9:07 a.m.

6. How many seconds are there in six days?

   A. 259,200 seconds

   B. 360,000 seconds

   C. 518,400 seconds

   D. 144,000 seconds

> Michael sold 4.5 gallons of root beer on Friday, 25 quarts on Saturday, and 55 pints on Sunday.

7. How many gallons of root beer did he sell on Saturday?

   A. 5 gal.

   B. 5.50 gal.

   C. 6 gal.

   D. 6.25 gal.

8. How many pints of root beer he did sell on Friday?

   A. 36 pints.

   B. 32 pints

   C. 48 pints.

   D. 35 pints.

9. How many gallons he did sell in 3 days?

   A. 16.225 gal

   B. 17.625 gal.

   C. 18.125 gal.

   D. 18.50 gal.

10. Dana went to a party. It started at 7:40 p.m. and ended at 11:25 p.m. How long was the party?

   A. Four hours and 5 minutes.

   B. Three hours and 30 minutes.

   C. Three hours and 45 minutes

   D. Four hours and 15 minutes.

Answer Key:

1) D	4) A	7) D	10) C
2) C	5) B	8) A	
3) B	6) C	9) B	

Answer the following reflection questions and feel free to discuss your responses with your teacher or a classmate.

1- What math ideas and principles did you learn in this chapter?

2- What new math concepts did you learn?

3- What procedures or methods did you practice in this chapter?

4- What aspect of this chapter is still not 100% clear for you?

5- What else do you want your teacher to know?

CHAPTER 5:
DATA ANALYSIS AND STATISTICS

Lesson 1: Interpret simple data sets, bar graphs, line graphs, and histograms.

A line graph is used to display data that changes continuously over time. It allows us to see overall trends such as an increase or decrease in data over time.

Line Graph

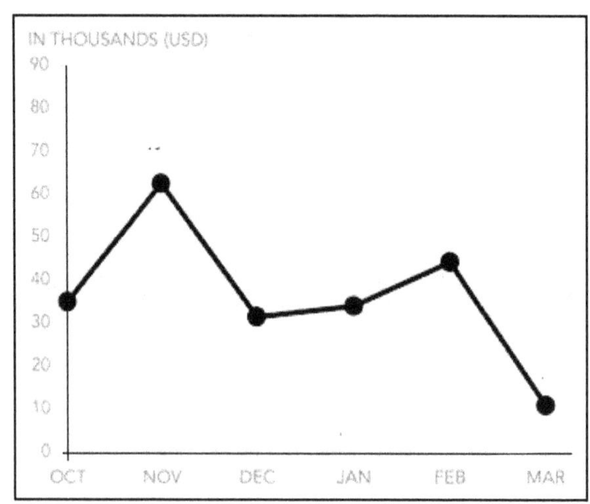

A **bar graph** is a graphical representation of data using bars or strips. It is used to compare and contrast different types of data or distinct categories of data.

Bar graph

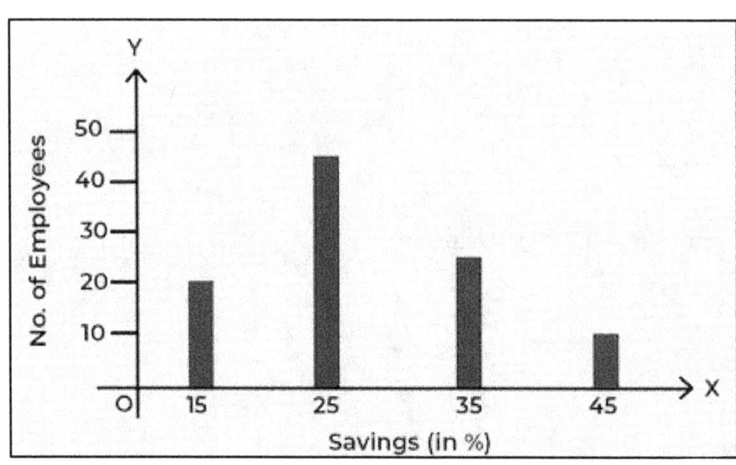

A histogram is a graphical display of data using bars of different heights. It is similar to a bar graph, but it groups numbers into ranges.

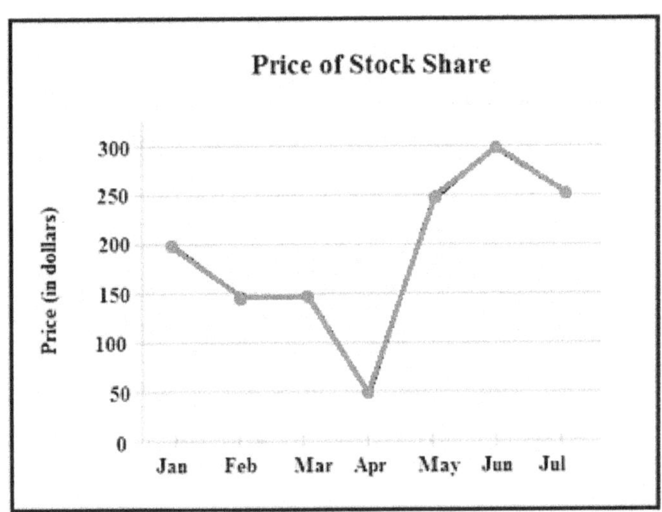

Example 1:

According to the following linear graph, in which month was the lowest price registered?

Solution:

Notice that the lowest price was $50. This value corresponds to **April.**

Practice Exercises

The following linear graph shows the number of people in a museum at different times on a particular day.

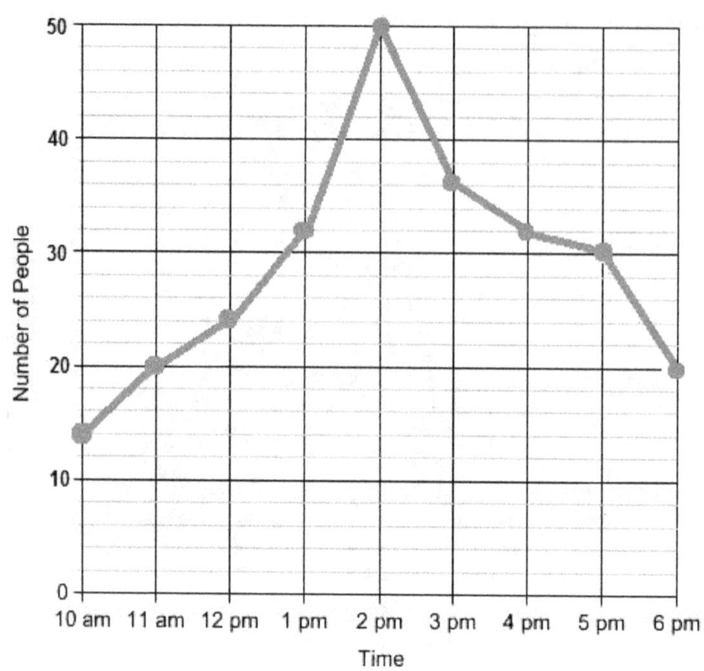

1. What was the greatest number of people in the museum?

    A. 40   C. 50

    B. 45   D. 30

2. What was the least number of people in the store?

    A. 12   C. 10

    B. 15   D. 14

3. How many people were in the museum when it opened?

    A. 14   C. 2

    B. 8   D. 10

4. How many people were in the museum at 4 pm?

    A. 30   C. 34

    B. 32   D. 35

5. Around what time did the number of people start decreasing?

    A. At 6 pm   C. Before 1 pm

    B. After 4 pm   D. After 2 pm

The following histogram shows the height of students within a class:

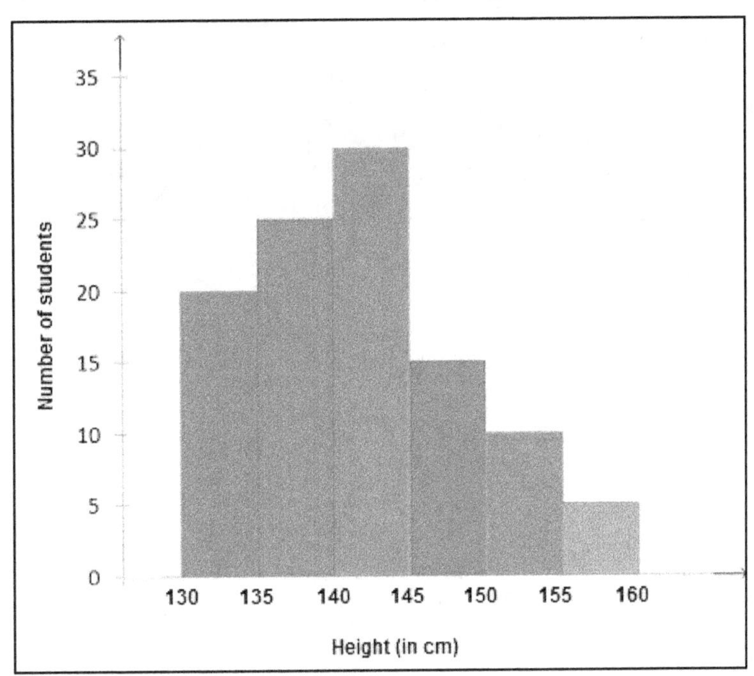

6. How many students have a height greater than 140 cm but less than 150 cm?

    A. 15   C. 45

    B. 30   D. 20

7. How many students have a height of more than 150 cm?

    A. 25   C. 30

    B. 45   D. 15

8. How many students have a height of less than 140 cm?

    A. 40   C. 30

    B. 45   D. 25

9. What is the total number of students?

    A. 100   C. 90

    B. 105   D. 95

10. How many students are there in the tallest range?

    A. 5   C. 10

    B. 15   D. 20

1) C   4) B   7) D   10) A

2) D   5) D   8) B

3) A   6) C   9) B

Lesson 2: Solve one- and two-step problems using bar graphs.

A **bar graph** is a chart that uses bars to compare different categories of data. The bars can be horizontal or vertical and the graph has two axes. One axis shows the types of categories being compared and the other has numerical values that represent the values of the data.

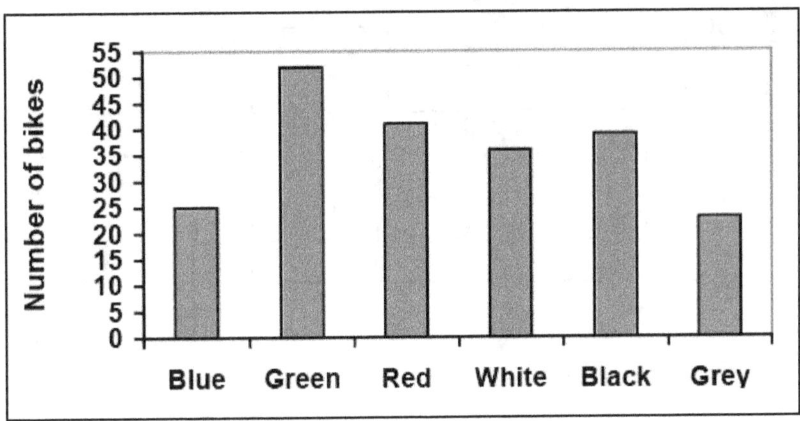

Example 1:

The following bar graph shows the number of cookies made by each chef. What is the lowest value in the data set?

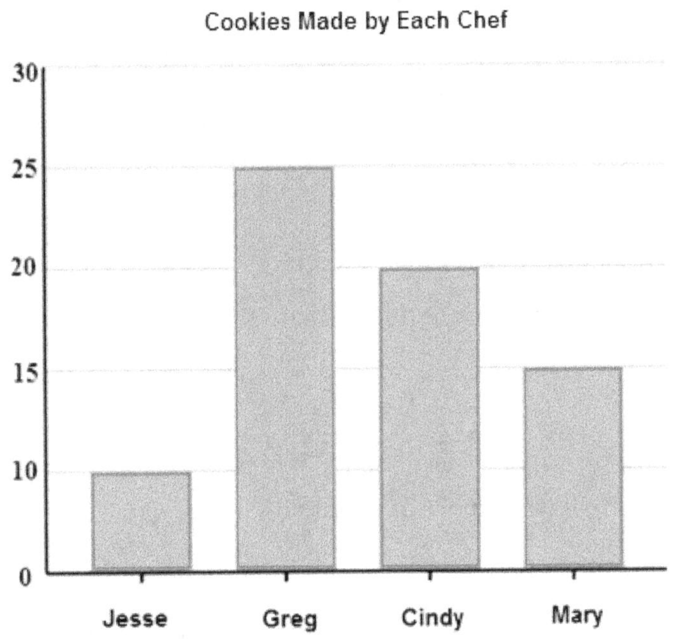

Solution:

Notice that the shortest bar has a height of 10 (Jesse). Therefore, the lowest value is **10.**

Practice Exercises

The following bar graph shows the number of students in different clubs.

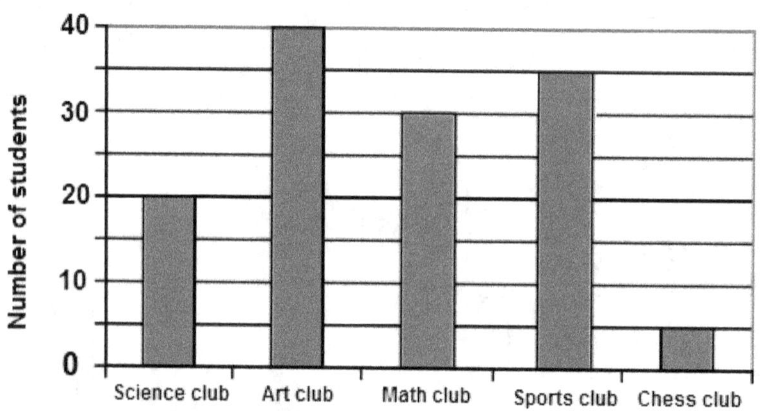

1. How many students are in the Sports Club?

   A. 30

   B. 32

   C. 35

   D. 40

2. Which is the most popular club?

   A. Sports Club

   B. Math Club

   C. Art Club

   D. Science Club

3. How many more students are there in the Sports Club than the Chess Club?

   A. 30

   B. 20

   C. 35

   D. 15

4. What is the total number of students in the clubs?

   A. 90

   B. 120

   C. 125

   D. 130

5. Which club is liked by the least number of people?

   A. Science Club

   B. Chess Club

   C. Math Club

   D. Sports Club

James was interested in the number of days it would take for an order from Amazon to arrive at his door. The following graph shows the data he collected.

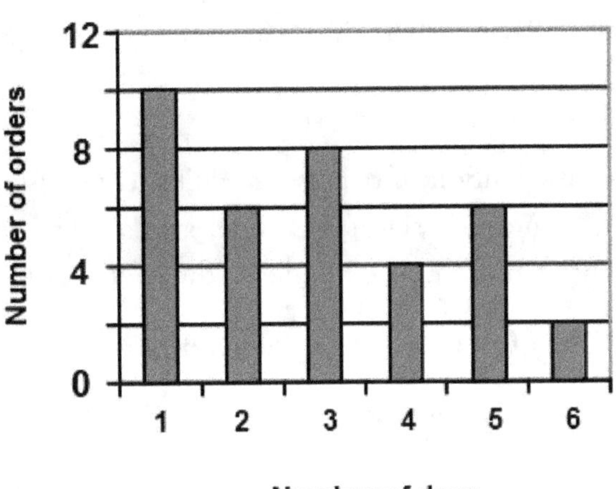

6. How many orders arrived in one day?

   A. 8   C. 4

   B. 10   D. 12

7. How many orders arrived in four days?

   A. 6   C. 5

   B. 2   D. 4

8. How many orders arrived in less than three days?

   A. 12   C. 14

   B. 16   D. 8

9. What was the lowest number of orders registered?

   A. 0   C. 2

   B. 1   D. 4

10. How many orders were registered?

   A. 36   C. 38

   B. 32   D. 26

Answer Key:

1) C	4) D	7) D	10) A
2) C	5) B	8) B	
3) A	6) B	9) C	

Lesson 3: Understand statistical variability concepts like "center" and "spread."

One important aspect of the distribution of data is where its center is located. The **mean, median, and mode** are measures of the center of a set of data. They are called **measures of central tendency**. They provide a single value that represents the middle or the center of the distribution.

The mean (or average) is the sum of all the values in the set divided by the number of values in the set.

The median is the **middle** of a sorted list of numbers.

The **mode** is the most common number that appears in a data set.

The mean has one main disadvantage: it is easily affected by **outliers**. **Outliers** are data points that are much higher or lower than the others. Therefore, the mean can change a lot because of an outlier. It is the only measure of central tendency that is always affected by an outlier.

In a symmetrical distribution of data, the mean is the same number as the median and mode.

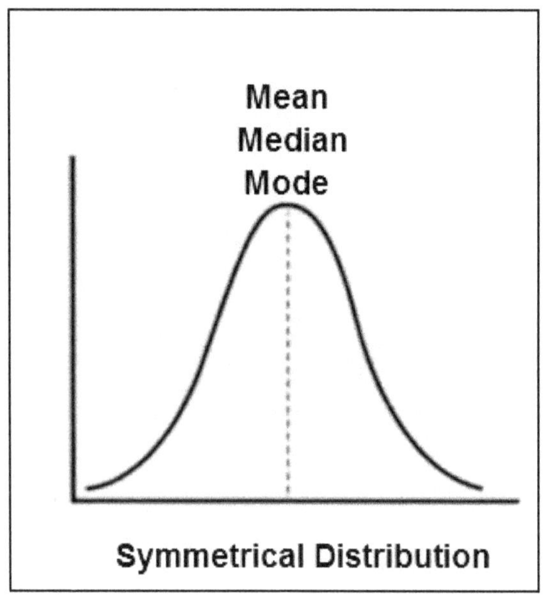

Measures of spread describe how similar or varied a set of values are. To find the spread in a data set, we must identify the **range**, which is the difference between the highest and lowest values. A larger range indicates a greater spread in the data.

Example 1:

Find the mean, median, mode, and range of the following data set: 18, 13, 15, 24, 10, 11, 17, 20

Solution:

Mean: To calculate the mean, add all of the numbers in the set and then divide the sum by how many numbers there are. (There are 8 numbers.)

$$Mean = \frac{18 + 13 + 15 + 24 + 10 + 11 + 17 + 20}{8} = \frac{128}{8} = 16$$

Then, the mean is **16.**

Median: To calculate the median, first place all of the values in numerical order.

$$10, 11, 13, 15, 17, 18, 20, 24$$

If we have an odd number of values, the median is the middle number. However, if we have an even number of values, the median will be the mean of the two values in the center of the data set.

In this case, we have an even number of values, so we must calculate the mean of the two central values:

$$10, 11, 13, \mathbf{15, 17}, 18, 20, 24$$

$$Mean = \frac{15 + 17}{2} = \frac{32}{2} = 16$$

Then, the median is **16.**

Mode: The mode is the value that occurs most often. Notice that no number appears more than once. Thus, there is no mode.

Range: To calculate the range of the data set, subtract the lowest value from the highest value.

$$Range = 24 - 10 = \mathbf{14}$$

Practice Exercises

Byron visited six gas stations. The price of gas at each station is shown in the following table.

Station	Gas Price (per gallon)
A	$2.95
B	$3.05
C	$2.77
D	$2.39
E	$2.84
F	$3.36

1. What is the mean price of gas?

   A. $2.89

   B. $2.45

   C. $2.39

   D. $2.77

2. What is the median price of gas?

   A. $2.84

   B. $2.95

   C. $2.90

   D. $3.05

3. What is the mode of the price of gas?

   A. $3.05

   B. $2.77

   C. $2.39

   D. There is no mode.

4. What is the price range of gas?

   A. $3.36

   B. $0.97

   C. $2.39

   D. $1.07

5. Which of the following is the measure for the center of a distribution of data?

   A. Range

   B. Outlier

   C. Variable

   D. Median

> The monthly salaries of five people are as follows:
> $8,950, $7,990, $4,475, $8,988, $9,123

6. What measure of central tendency will most accurately describe the data?

   A. Median

   B. Range

   C. Mean

   D. Mode

7. What is the mean of the monthly salaries?

   A. $7,990

   B. $8,201.55

   C. $7,905.20

   D. $9,100

8. What is the median of the monthly salaries?

   A. $4,475

   B. $7,990

   C. $8,988

   D. $8,950

9. What is the range of the monthly salaries?

    A. $4,475   C. $7,990

    B. $4,648   D. $5,600

10. What is the outlier in the data set, if one exists?

    A. $7,990   C. $4,475

    B. $9,123   D. There is no outlier.

Answer Key:

1) A	4) B	7) C	10) C
2) C	5) D	8) D	
3) D	6) A	9) B	

Answer the following reflection questions and feel free to discuss your responses with your teacher or a classmate.

1- What math ideas and principles did you learn in this chapter?

2- What new math concepts did you learn?

3- What procedures or methods did you practice in this chapter?

4- What aspect of this chapter is still not 100% clear for you?

5- What else do you want your teacher to know?

CHAPTER 6:
PURE MATHEMATICS

Word Problems

1. A software company has 26,507 customers. How do you write this number in words?

2. A 4-digit number has a 5 in the thousands place, an 8 in the tens place, and 0s elsewhere. What is the number?

3. Rick bought 83 feet of electrical wire for $215.80. Find the price per foot.

4. How many quarters are there in $30?

5. In an exam, Linda gets twice the score as Joe. Two times Linda's score and four times Joe's score make 344. What is the Joe's score?

6. The perimeter of the following triangle is 30 feet. What is the area of the triangle?

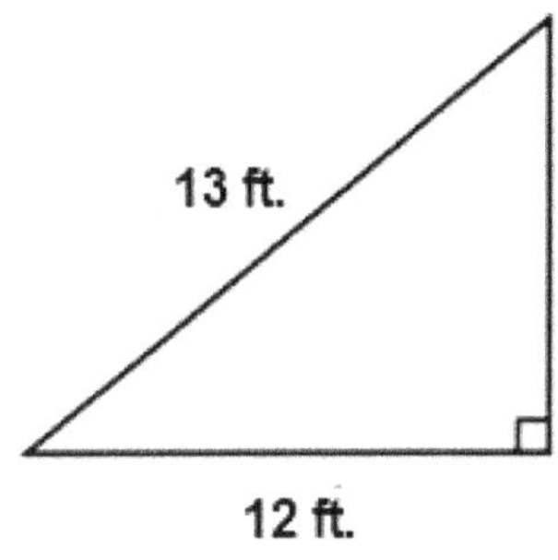

7. A family plants a rectangular 9.5-yard by 14.2-yard organic garden. How many yards of fencing do they need to put around the garden?

8. Susan left the library at 4:43 pm and arrived home 1,920 seconds later. What time did she arrive home?

9. In three tests, Tom scored 81, 88, and 93. What does he have to score on the fourth test to average a 90 in the class?

10. Melissa earns $8.50 per hour at her babysitting job. On Saturday, she worked 6.5 hours. What were her total earnings for the day?

1) Twenty-six thousand five hundred seven

2) 5,080

3) $2.60 per foot

4) 120

5) 43

6) 30 ft^2

7) 47.4 yd.

8) 5:15 pm.

9) 98

10) $55.25

Practice Problems

1. A plumber needs to replace 34.40 feet of pipe in a home. He has 26.50 feet in his truck. How much more pipe does he need?

   A. 60.90 ft.

   B. 7.90 ft.

   C. 8.60 ft.

   D. 17.10 ft.

2. How do you write 3.09 in words?

   A. Three and nine hundred

   B. Three and nine hundredths

   C. Three and nine thousands

   D. Three and nine thousandths

3. John read 35 pages of a book in 50 minutes. How many pages should he be able to read in one hour?

   A. 70

   B. 86

   C. 84

   D. 75

Look at the following table:

Pants	Total Cost	Number of Pants	Unit Rate (Dollars per Pair of Pants)
Store 1	$320	A	$32 per pair
Store 2	$450	12	B
Store 3	$280	9	$31.11 per pair

4. What is A?

   A. 15

   B. 12

   C. 8

   D. 10

5. What is B?

   A. $37.50 per pair

   B. $38.15 per pair

   C. $30 per pair

   D. $33.50 per pair

6. What is the best deal?

   A. Store 2

   B. Store 1

   C. Store 3

7. What is N?

$$12 (8 + 15) = 96 + N$$

   A. 27   C. 150

   B. 180   D. 37

8. What is the area of a circle whose radius is 1/2 feet? (Use $\pi = 3.14$)

   A. 3.14 ft^2 C. 2.56 ft^2

   B. 1.57 ft^2 D. 0.785 ft^2

9. Which unit is most appropriate for describing the length of a bulldozer?

   A. Kilometers   C. Feet

   B. Hand span   D. Millimeters

10. The number of gallons of milk sold in eight days is 5.6, 5.1, 5.6, 5.6, 5.6, 5.6, 5.6, and 5.6. Which of the following is true?

   A. The mean is 5.6

   B. The mode is greater than the median.

   C. There are three outliers in the data set.

   D. The median is the same number as the mode.

Answer Key

1) B	4) D	7) B	10) D
2) B	5) A	8) D	
3) B	6) C	9) C	

<h1 style="text-align:center">PRACTICE TEST 1</h1>

Read each question and choose the best answer. You have 60 minutes to answer 36 questions.

1. Which is another way to show 23,709?

   A. 23,000 + 79

   B. 20,000 + 370 + 9

   C. 20,000 + 3,000 + 700 + 9

   D. 20,000 + 3,000 + 700 + 90

2. In the number 342,027, what digit is in the thousands place?

   A. 2

   B. 4

   C. 0

   D. 3

3. A bakery's budget for utilities for an entire year is $15,048. What would the monthly budget for utilities be?

   A. $1,254

   B. $1,300

   C. $1,505

   D. $1,356

4. If a server makes $18.45 per hour, including tips, how much would he make after working a 50-hour week?

   A. $68.45

   B. $822.50

   C. $922.50

   D. $584.80

5. Which of the following is a unit rate?

   A. 4 books/ 3 minutes

   B. 3.45 inches

   C. 12.4 ft^2

   D. $12.50 per hour

Look at the following table:

Item	Cost	Amount
A	$495	15 pounds
B	$576	18 pounds
C	$350	10 pounds

6. What is the unit price of Item A in dollar per pound?

    A. $35 per pound   C. $33 per pound

    B. $34 per pound   D. $30 per pound

7. What is the cost of 25 pounds of item B?

    A. $800   C. $825

    B. $600   D. $850

8. What is the best deal?

    A. Item C   C. Item A

    B. Item B

Look at the following receipt:

9. How many items were purchased?

    A. 5   C. 7

    B. 6   D. 4

10. What is the subtotal amount?

A. $535.85

C. $49.89

B. $533.89

D. $12.98

11. What is the total amount?

A. $595.74

C. $585.74

B. $5355.85

D. $577.04

12. Which item is the cheapest?

A. Dave Conditioner

C. Bluetooth

B. Dave Shampoo

D. Discount coupon

13. What is the cost of six Dave Shampoo bottles?

A. $53.94

C. $97.18

B. $18.98

D. $77.88

14. Which of the following is equal to 30 (40 − 13)?

A. 1200 − 390

C. 700 − 390

B. 30 (13 − 40)

D. 40 (30 − 13)

15. What is the missing number?

$$245 \times 98 = 98 \times \, ?$$

A. 98

C. 343

B. 245

D. 200

16. Eight friends share the cost of a restaurant bill. The bill is $368. How much does each person pay?

A. $50

C. $52

B. $45

D. $46

17. Solve the following equation:

$$5x - 15 = 35$$

A. x = 10

B. x = 4

C. x = 5

D. x = 12

18. On a test, the highest grade is 29 points higher than the lowest grade. The sum of the two grades is 151. What is the lowest grade?

A. 90

B. 61

C. 72

D. 59

19. Five times a number is equal to 200. What is the equation that represents this situation?

A. x + 5 = 200

B. x = 5 (200)

C. 5x = 200

D. 200x = 5

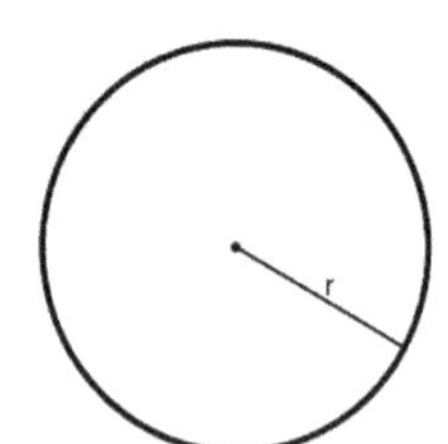

The formula for the circumference of the circle is

$$C = 2\pi r$$

C is the circumference of the circle and r is

20. What is the circumference of a circle with a radius of 0.5 feet? (Use $\pi = 3.14$)

A. 1 ft.

B. 6.28 ft.

C. 3.14 ft.

D. 1.57 ft.

21. If the circumference of the circle is 62.8 inches, what is the radius? (Use $\pi = 3.14$)

A. 20 in.

B. 5 in.

C. 15 in.

D. 10 in.

22. If the circumference of the circle is 50.24 inches, what is the area? (Use $\pi = 3.14$)

A. 100.48 in^2

B. 150.36 in^2

C. 200.96 in^2

D. 150.72 in^2

23. What is the area of the following shape?

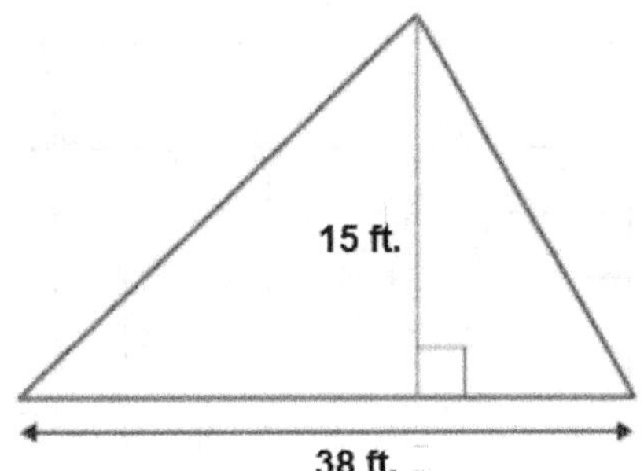

A. 285 ft²   C. 236 ft²

B. 570 ft²   D. 108 ft²

24. Which of the following is a metric unit of length?

A. Feet span   C. Yard

B. Spoon   D. Kilometer

25. Mrs. Woods is measuring out a small quantity of medicine to put in a capsule. What units will she use to measure the mass of the medicine?

A. Gallon   C. Kilogram

B. Milligram   D. Pound

26. Which of the following is true?

A. A kilometer is smaller than a meter.

B. A pencil is a metric unit of length.

C. A spoon is a metric unit of volume.

D. A barrel is a non-standard unit of volume.

27. How many gallons are there in 500 quarts?

A. 250 gal.   C. 125 gal.

B. 150 gal.   D. 175 gal.

28. Gina purchased 5.75 meters of rope, and Larry purchased 4,000 centimeters of rope. What is the total length in meters of ropes they both purchased?

A. 9.75 m.   C. 45.75 m.

B. 6.15 m   D. 10.75 m.

A student recorded how much time he spent reading during five days. The following table shows the results.

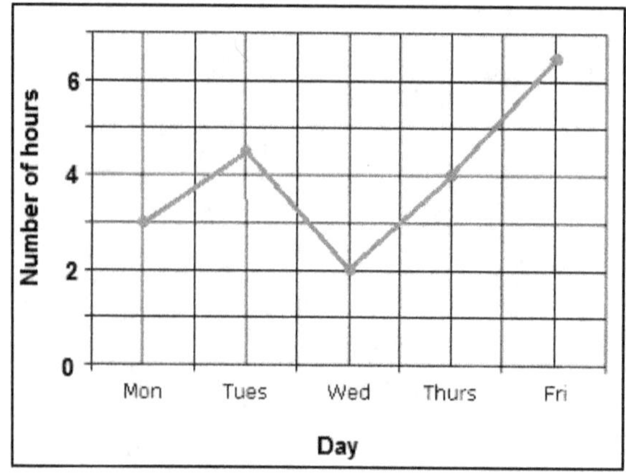

29. How many hours did he spend reading on Thursday?

A. 3 hours

B. 4 hours

C. 5 hours

D. 2 hours

30. About how many hours did he spend reading on Friday?

A. 7 hours.

B. 6.9 hours

C. 6 hours

D. 6.5 hours

31. Which day did he spend three hours reading?

A. Friday

B. Wednesday

C. Monday

D. Tuesday

32. What is the total number of hours he spent reading during the five days?

A. 19

B. 15

C. 20

D. 18

> The prices of the same brand of peanut butter at
> 6 stores were $4.15, $3.99,
>
> $4.05, $3.89, $3.90, and $3.79

33. What is the mean price?

A. $3.91

B. $3.96

C. $3.90

D. $4.05

34. What is the median price? (Round your answer to decimals.)

A. $3.95

C. $3.99

B. $3.90

D. $3.93

35. What is the outlier in the data set, if one exists?

A. $4.15

C. $3.99

B. $4.05

D. There is no outlier.

36. What is the range of the data set?

A. 0.72

C. 0.10

B. 0.36

D. 0.57

REFLECTION ON LEARNING

Answer the following reflection questions and discuss your responses with your teacher or a classmate.

1- How do you feel about your performance on the test?

2- Which types of questions were difficult for you?

3- How do you feel about your time management strategies?

4- What specific things do you want to do differently next time? List them.

5- What math functions or content areas do you want to review? List them.

6- What else do you want your teacher to know?

Answer Key:

1) C	10) A	19) C	28) C
2) A	11) C	20) C	29) B
3) A	12) A	21) D	30) D
4) C	13) D	22) C	31) C
5) D	14) A	23) A	32) C
6) C	15) B	24) D	33) B
7) A	16) D	25) B	34) A
8) B	17) A	26) D	35) D
9) D	18) B	27) C	36) B

PRACTICE TEST 2

Read each question and choose the best answer. You have 60 minutes to answer 36 questions.

1. In the number 834,568, what digit is in the ten thousands place?

   A. 8

   B. 4

   C. 6

   D. 3

2. Rebecca walks 0.567 miles to and from school each day. What is this number in word form?

   A. Five hundred sixty-seven thousand

   B. Five hundred six seven thousandths

   C. Six hundred fifty-seven thousandths

   D. Five hundred sixty-seven thousandths

3. James has $789.67 in his checking account. How much does he have in his account after he makes a deposit of $288.95 and a withdrawal of $177.32?

   A. $901.30

   B. $612.35

   C. $863.11

   D. $550.82

4. Cindy studied a total of 36.5 hours over five days. On average, how many hours did she study each day?

   A. 5.6 hours

   B. 6.7 hours

   C. 7.3 hours

   D. 7.2 hours

5. Compute

$$\frac{1}{5}\left(\frac{3}{4}+\frac{1}{2}\right)$$

   A. $\dfrac{1}{2}$

   B. $\dfrac{1}{4}$

   C. $\dfrac{5}{6}$

   D. $\dfrac{2}{3}$

6. Mark can mow a lawn that measures 1,200 square feet in 1.5 hours. At that rate, how long would it take him to mow a lawn 5,400 square feet?

   A. 5.50 hours

   B. 6.20 hours

   C. 7 hours

   D. 6.75 hours

Look at the following table:

Item	Cost
Movie ticket	$9.50
Popcorn	$5.50
Soda	$4.25
Candy	$2.99

7. What is the total cost of three movie tickets and four popcorn?

A. $50.50

B. $48.75

C. $51.25

D. $60.00

8. If Jake spent $57 on Movie tickets, how many tickets did he buy?

A. 5

B. 6

C. 8

D. 7

9. Which item is the cheapest?

A. Movie ticket

B. Soda

C. Candy

D. Popcorn

Look at the following grocery receipt:

08-25-2013	5:14 p.m.

Fred's Superstore

2% milk 4 L	$4.76
orange juice 1L	$3.79
bread	$2.49
eggs 1 doz	$3.69
bananas 3 Kg	$1.83
cereal 400 g	$4.39
TOTAL	$20.95
CASH	$25.00
CHANGE	?

10. How many items were purchased?

   A. 5　　　　　　　　　　　　C. 4

   B. 7　　　　　　　　　　　　D. 6

11. How many eggs did the customer buy?

   A. 10　　　　　　　　　　　C. 6

   B. 12　　　　　　　　　　　D. 18

12. How much change did the customer get?

   A. $4.05　　　　　　　　　　C. $4.00

   B. $4.50　　　　　　　　　　D. $4.15

13. Which item is the most expensive?

   A. Bananas　　　　　　　　　C. Milk

   B. Cereal　　　　　　　　　　D. Orange juice

14. What is the cost of four liters of orange juice?

   A. $14.75　　　　　　　　　C. $15.16

   B. $4.76　　　　　　　　　　D. $13.85

15. What is the value of P?

$$12 \times 56 \times 31 = 56 \times P \times 12$$

   A. 31　　　　　　　　　　　C. 56

   B. 12　　　　　　　　　　　D. 12 x 31

16. Which expression is equivalent to 300 x 1,009?

   A. 3 (1000) + 9　　　　　　C. 1,009 x (3 + 100)

   B. 30 x 1,000 x 9　　　　　D. 300 (1000 + 9)

17. Solve the following equation:

$$36 + 8x = 100$$

A. 12

B. 8

C. 17

D. 9

18. Super's Bikes rents bikes for $12 plus $3 per hour. Scott paid $39 to rent a bike. For how many hours did he rent the bike?

    A. 8.5 hours

    B. 7 hours

    C. 9 hours

    D. 10.5 hours

Look at the following triangle:

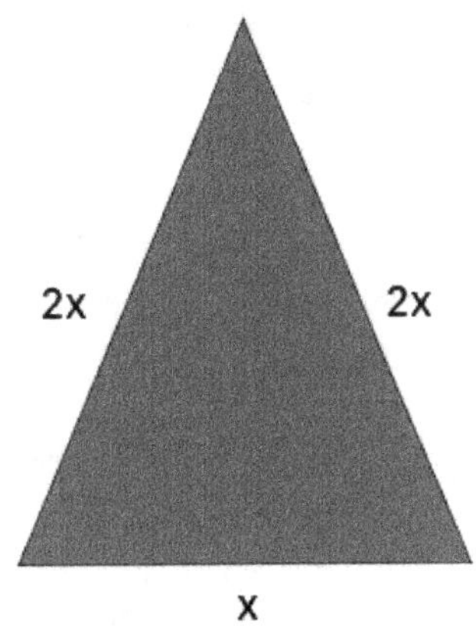

19. If the perimeter of the triangle is 60 inches, what is x?

    A. 15

    B. 18

    C. 16

    D. 12

20. If x = 4.5 inches, what is the perimeter of the triangle?

    A. 22.5 inches

    B. 13.5 inches

    C. 20 inches

    D. 18.5 inches

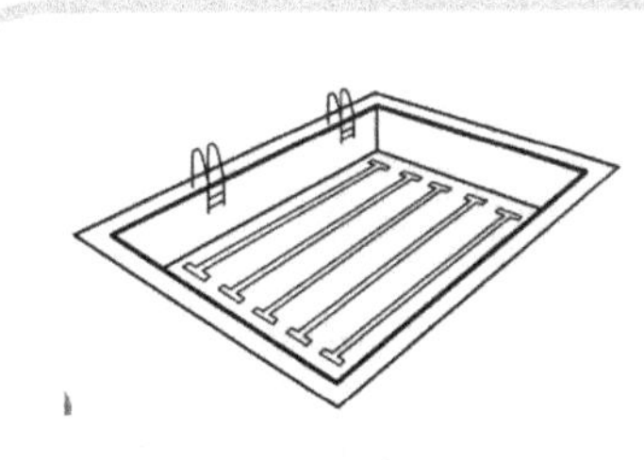

The perimeter of a rectangular pool is 82 feet. The length of the pool is 25 feet.

21. What is the width of the pool?

    A. 20 ft.

    B. 16 ft.

    C. 38 ft.

    D. 57 ft.

22. If the width and the length of the pool are 18 ft. and 27 ft. respectively, what is the perimeter of the pool?

    A. 45 ft.   C. 84 f.

    B. 76 ft.   D. 90 ft.

23. Which unit is most appropriate for describing the capacity of the pool?

    A. Meter   C. Gallons

    B. Cups   D. Milliliters

24. Which of the following would be a non-standard unit of weight?

    A. Kilogram   C. Foot

    B. Sandbag   D. Ounce

Andrew sleeps for 7 hours and 45 minutes in a day.

25. How many hours did he sleep in three days?

    A. 23.25 hours   C. 25 hours

    B. 21.50 hours   D. 24.75 hours

26. How many hours did he sleep in one week?

    A. 49 hours   C. 49.45 hours

    B. 51.45 hours   D. 54.25 hours

27. How many seconds did he sleep in a day?

    A. 26,500 seconds   C. 25,200 seconds

    B. 27,900 seconds   D. 26,820 seconds

28. Cassy has a fish tank with a capacity of 3 gallons. She uses a 3-cup bowl to fill the tank. How many bowls of water does she use?

    A. 24   C. 48

    B. 16   D. 12

The following bar graph shows the number of visitors to Rock Park from July to November.

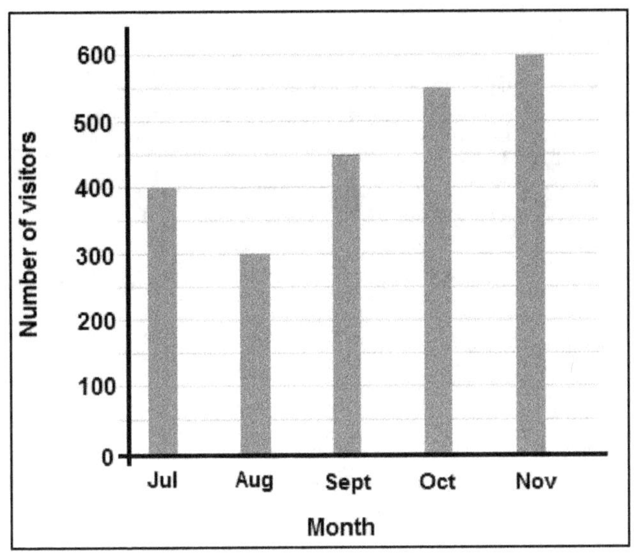

29. How many people visited Rock Park in October?

    A. 600   C. 500

    B. 550   D. 400

30. Which month were there 300 visitors to the Rock Park?

    A. July   C. September

    B. October   D. August

31. Which month was the maximum number of visitors recorded?

    A. November   C. July

    B. September   D. August

32. Which month was the minimum number of visitors registered?

    A. July   C. August

    B. October   D. September

> The times in minutes it took for 7 participants to solve
> a puzzle are as follows: 8.5, 9.3, 7.7, 8.8, 9.7, 7.9 and 9.3

33. What is the mean of the data set?

    A. 9.3   C. 8.74

    B. 8.95   D. 7.7

34. What is the median of the data set?

    A. 8.8

    B. 8.5

    C. 9.3

    D. 7.7

35. What is the mode of the data set?

    A. 8.5

    B. 9.3

    C. 8.8

    D. There is no mode.

36. What is the range of the data set?

    A. 1.6

    B. 0.5

    C. 7.7

    D. 2.3

REFLECTION ON LEARNING

Answer the following reflection questions and discuss your responses with your teacher or a classmate.

1- How do you feel about your performance on the test?

2- Which types of questions were difficult for you?

3- How do you feel about your time management strategies?

4- What specific things do you want to do differently next time? List them.

5- What math functions or content areas do you want to review? List them.

6- What else do you want your teacher to know?

1) D	10) D	19) D	28) B
2) D	11) B	20) A	29) B
3) A	12) A	21) B	30) D
4) C	13) C	22) D	31) A
5) B	14) C	23) C	32) C
6) D	15) A	24) B	33) C
7) A	16) D	25) A	34) A
8) B	17) B	26) D	35) B
9) C	18) C	27) B	36) A

About CBL

At CBL, we promote systematic solutions, learner-centered textbooks, and forward-thinking strategies in adult education, workforce development, and vocational training. Our diverse solutions and products are intricately designed to enrich students' learning experiences while making the job of busy, hard-working adult instructors easier.

CBL takes pride in publishing student-centered textbooks designed to prepare learners for CASAS, TABE 11&12, HiSET, and GED assessments and assist instructors in covering course curricula and standards with confidence.

Our publications also include teaching guides, test prep tools, and study guides that foster reflective learning, ensuring sustained engagement in active learning. Find our meticulously crafted textbooks on our book page (cbledu.com) or major platforms like Amazon, Barnes & Noble, and Ingram Spark.

CBL also guides adult education and workforce programs in establishing robust professional development programs—training, peer-mentoring, coaching, community of practices (CoPs), and instructional systems— fostering a culture of continuous improvement and contributing to higher learner retention and success rates. We also offer workshops and PD sessions for adult educators and classroom instructors.

If you have questions about instructional systems, textbooks, or student learning and retention, contact us today at teamcbl@cbledu.com or 410-960-4082.